SpringerBriefs in Molecular Science

For further volumes:
http://www.springer.com/series/8898

S. Sundaram · P. S. Raghavan

Chromium-VI Reagents: Synthetic Applications

 Springer

Prof. S. Sundaram
Department of Chemistry
Ramakrishna Mission
Vivekananda College
Mylapore
Chennai 600004
India

Prof. P. S. Raghavan
Department of Chemistry
Madras Christian College
Tambaram
Chennai 600059
India
e-mail: ps_raghavan@rediffmail.com

ISSN 2191-5407
ISBN 978-3-642-20816-4
DOI 10.1007/978-3-642-20817-1
Springer Heidelberg Dordrecht London New York

e-ISSN 2191-5415
e-ISBN 978-3-642-20817-1

Cover design: eStudio Calamar, Berlin/Figueres

Printed on acid-free paper

Springer is part of Springer Science+Business Media (www.springer.com)

Preface

Chromium (VI) reagents are versatile oxidants used frequently by both academic and industrial chemists. In the past three decades, particularly from 1975, a number of new oxidizing agents containing Cr(VI) species have been discovered and published in various professional journals. Such newly discovered Cr(VI) oxidants have been found to possess certain specificity and selectivity in their behaviour. All such interesting and useful informations from various international journals have been collected; the methods of preparation of these compounds and their applications in different spheres have been critically evaluated, and are presented in this book in such a manner that this would be an easy-to-use ready reconer for persons working in this field. This monograph is the ONLY COMPREHENSIVE SOURCE of information on as many as 40 new generation Cr(VI) reagents, in our opinion, this publication will be useful for post-graduate chemistry students of all colleges and universities and chemists working in the academic and industrial research laboratories world wide. The information presented in this book spans as many as 284 original research papers published in various international professional journals and, only very significant and specific data on each of the reagents considered are presented in this book. Though the main attention is on the synthetic applications of these reagents, their utility in the kinetics and biological application is also significant and are outlined in Chaps. 4 and 6 respectively. However for more details, the readers are suggested to go through the original papers cited in the reference. The authors wish to acknowledge the help rendered by many, including some of their students in collecting research papers and their colleagues for many useful discussions and suggestions. The authors also request the readers who may come across any other reagents of this family which are left out by oversight to provide relevant information so that they can be included in the next edition. We would also be glad and thankful to receive any helpful suggestions and comments from the readers.

Chennai, April 2011   S. Sundaram
                                                             P. S. Raghavan

Contents

Abbreviations

Abbreviations used in the text and the corresponding name and structure of the reagents

Abbreviation	Compound	Formula/Structure
BIDC	Benzimidazolium dichromate	$Cr_2O_7^{2-}$
BIFC	Benzimidazolium fluorochromate	$FCrO_3^-$
BTEACC	Benztriethylammonium chlorochromate	$C_6H_5CH_2-(C_2H_5)_3N^+ClCrO_3^-$
BPCC	Bipyridinium chlorochromate	$ClCrO_3^-$
3CPDC	3-Carboxypyridinium Dichromate (Nicotinium dichromate)	$Cr_2O_7^{2-}$
CTADC	Cetyltrimethylammonium dichromate	$[CH_3-(CH_2)_{14}-CH_2-N^+-(CH_3)_3]_2\ Cr_2O_7^-$

(continued)

(continued)

Abbreviation	Compound	Formula/Structure
3CPCC	3-Carboxypyridinium chlorochromate (Nicotinium chlorochromate)	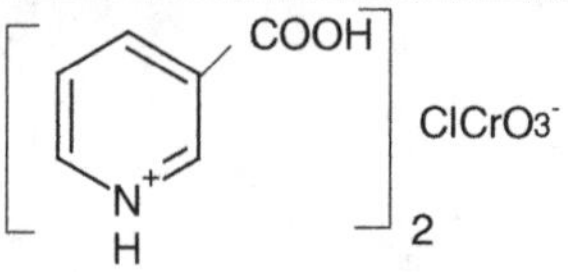
DCPCC	2,6-Dicarboxypyridinium chlorochromate	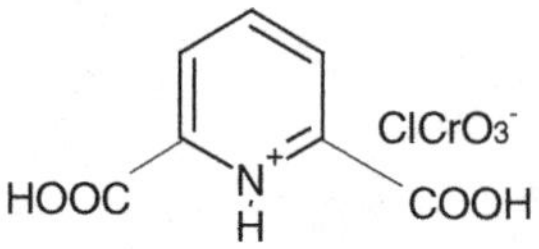
DCPFC	2,6-Dicarboxypyridinium fluorochromate	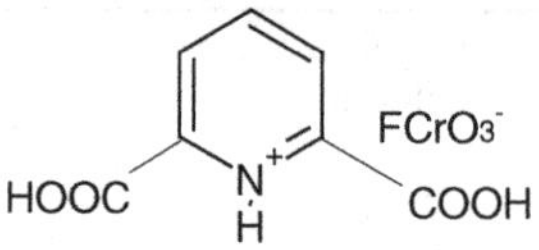
DMAPCC	4-N,N-Dimethylaminopyridinium Chlorochromate	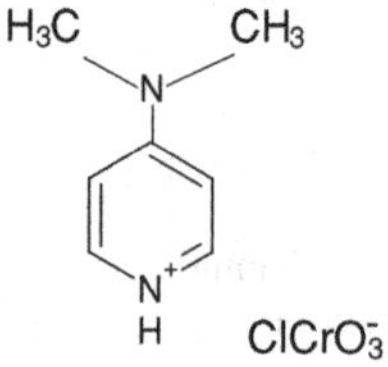
GCC	Guanidinium chlorochromate	$(NH_2)_2C(NH)^+ClCrO_3^-$
ICC	Imidazolium chlorochromate	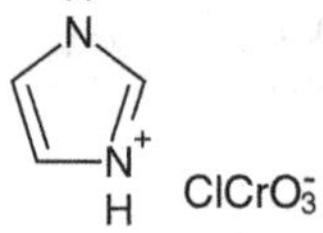

(continued)

(continued)

Abbreviation	Compound	Formula/Structure
IFC	Imidazolium fluorochromate	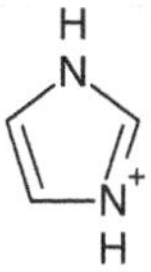 $FCrO_3^-$
MICC	1-Methylimidazolium chlorochromate	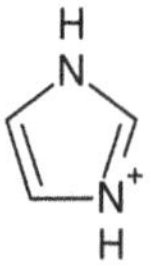 $ClCrO_3^-$
NCC	Naphthyridium chlorochromate	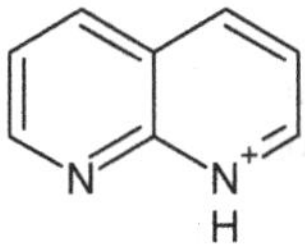 $ClCrO_3^-$
PDC	Pyridinium dichromate	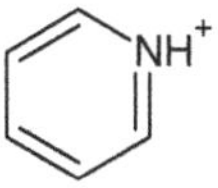 $Cr_2O_7^{--}$
PCC	Pyridinium chlorochromate	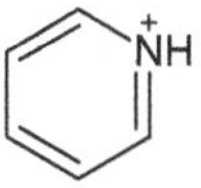 $ClCrO_3^-$
PFC	Pyridinium fluorochromate	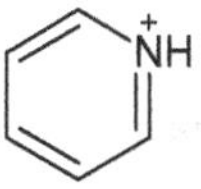 CrO_3F-
PBC	Pyridinium bromochromate	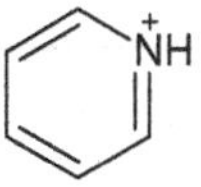 CrO_3Br^-

(continued)

(continued)

Abbreviation	Compound	Formula/Structure
PzCC	Pyrazinium chlorochromate	$ClCrO_3^-$
PVPDC	Polyvinylpyridinium dichromate	$Cr_2O_7^{2-}$
QCC	Quinolinium chlorochromate	$ClCrO_3^-$
QFC	Quinolinium fluorochromate	CrO_3F^-
QBC	Quinolinium bromochromate	CrO_3Br^-
TPSDC	Tetrakispyridiniumsilver dichromate	$[Ag(Py)_2]_2\ Cr_2O_7$
TBACC	Tetrabutylammonium chlorochromate	$[C_4H_9]_4N^+ClCrO_3^-$
TMSCC	Trimethylsilyl chlrochromate	$(CH_3)_3SiH\ ClCrO_3$

Chapter 1
Introduction

Oxidation is an important process in many natural events, including our life cycle. Oxidation is defined as the addition of oxygen, loss of hydrogen or loss of electrons. Thus the molecular oxygen itself or a substance which can release oxygen on decomposition, such as peroxides, can act as oxidizing agents. On the other hand, metals being highly electropositive, they tend to lose electrons easily and thus any metal atom which can exist in more than one valency can function as an oxidant under suitable conditions. Consequently oxidants may be classified as metal oxidants and non-metal oxidants.

Among the non-metal oxidants, molecular oxygen, ozone, peroxides, halogens, nitric acid, nitrous acid, etc. are commonly used substances. Usually, molecular oxygen requires relatively elevated temperature or light or a catalyst for its performance. However, *in vivo* systems intelligently make the oxygen react exactly at the target site by transporting it in the form of a complex with molecules like haemoglobin and myoglobin and delivering it in high local concentration at the site of the reaction! Ozone is a powerful oxidant and the well known ozonolysis of alkenes, which passes through the formation of a five- member malozanide is highly useful in determining the position of unsaturation in the alkene. Halogens are also good oxidants and their oxidizing ability follows the order $F_2 > Cl_2 > Br_2 > I_2$. The oxidizing abilities of nitric acid, nitrous acid, periodic acid and other non-metal oxidants are well understood and discussed in detail in many standard inorganic chemistry text books.

Among the metal-ion oxidants, Ag^+ and Cu^{2+} are relatively mild oxidants and are best suited for oxidizing aldehyde groups to carboxylic acid in the presence of other oxidisable functions such as alcohols. For example, Tollen's reagent, $Ag(NH_3)^{2+}$, Fehling's and the Benedict's solutions consisting of Cu^{2+} ions in presence of tartrate or citrate respectively are well known reagents. Metal ions such as Ce^{4+}, Cr^{6+}, Mn^{7+}, V^{5+} and Os^{8+} are powerful oxidants commonly employed to oxidize several organic functional groups. All of them are good oxidants specifically in acid medium while Mn^{7+} can do so even in basic medium.

S. Sundaram and P. S. Raghavan, *Chromium-VI Reagents: Synthetic Applications*,
SpringerBriefs in Molecular Science, DOI: 10.1007/978-3-642-20817-1_1,
© P. S. Raghavan 2011

However, as this book deals mainly with Cr(VI) oxidants only, a detailed account of various oxidizing agents containing hexavalent chromium ion is given in the following chapters.

1.1 Chromium: The Element

Chromium is a steel-gray, lustrous and hard metal. It was discovered by Louis Nicholas Vauquelin in 1798 from the mineral crocolite, which consists of lead chromate. The name chromium is derived from the Greek word *chroma*, which means color, owing to the different colors produced by its compounds. Chromium is the 21st most abundant element found on the Earth's crust. However, it never occurs as a free metal and is extracted from the mines as chromite ore. Chromium occurs in many oxidation states +1, +2, +3, +4, +5, +6, of which +3 and +6 are most common, while +1, +4 and +5 are the rare oxidation states. Chromium has found a wide range of applications, mainly due to its hardness and resistance to corrosion. It is also known for its remarkable magnetic property. It is mainly used in the manufacture of stainless steel, along with nickel, as it prevents corrosion and discoloration of steel. Today, chromium is a very important alloying material for steel.

Acidic chromate or dichromate solutions are also used for surface coating. This is usually done with the help of electroplating technique, in which a thin layer of chromium is usually deposited on the surface of metals. However, for imparting wear resistance quality, a thick layer is required to be deposited. Alternatively the chromate conversion coating process can be employed in which chromates are used to deposit a protective layer on certain metals like aluminium, cadmium, zinc, silver and magnesium.

Salts of chromium are used for their toxic properties in preserving wood from decay and damage caused by fungi, insects and termites. Chromium (III) salts are also used in leather tanning. The high melting point and resistance to heat makes chromite and chromium oxide an ideal refractory material. They have found applications in blast furnaces, cement kilns and metal casting. Besides, many compounds of chromium are used as catalysts. Chromium (IV) oxide is used to produce magnetic tapes, which are used in audio tapes and cassettes.

Chromium is also used in pigment industries. Chrome yellow, made of lead chromate was widely used as a pigment in the past, but due to environmental issues, its use has significantly declined as it contains lead, which is a toxic material. Other pigments of chromium include chrome red, chrome oxide green and chrome green, which is a mixture of chrome yellow and Prussian blue. Chromium oxide is used for imparting greenish color to glass. Besides, some precious stones also owe their attractive colour and appearance to the presence of chromium. Emerald is a form of beryl (beryllium aluminium silicate) which is green because of the inclusion of a little chromium into the beryl crystal lattice in the place of some of the aluminium ions. Similarly, traces of chromium incorporated into the crystal lattice of corundum (crystalline aluminium oxide, Al_2O_3) as a replacement for some of the Al^{3+} ions results in another highly coloured gem stone, the red ruby. Chromium oxide is also used in manufacturing synthetic rubies. Besides, it is also required as a trace element in the human body.

1.2 Chromium (VI) as Oxidant

Hexavalent chromium is a well known and versatile oxidizing agent. The acid dichromate, chromyl chloride and chromyl acetate were known for about a century and their reactions have been well documented [1–11]. They oxidize the activated C–H bonds emerging from an aromatic ring so completely that they usually convert any alkyl benzene to benzoic acid. Cr(VI) reagents are so powerful that they can also oxidize alkenes and alkynes, breaking the C–C bonds. Following the introduction of pyridinium chlorochromate by Corey and co-workers [12] in 1975, a myriad of modified versions of Cr(VI) oxidants has emerged on the horizon of organic synthesis. The concurrent appearance of two reviews [13, 14] in the year 1982, upholds the bursting profile of these novel Cr(VI) reagents developed within a short span of about seven years. One of these reviews [14] is exclusively devoted to pyridinium chlorochromate and it emphasizes the growing utility of this reagent. During the last one and half decades of the twentieth century, more than thirty new Cr(VI) reagents have been introduced and varied claims made about their utility. An overall perception of this fast developing family of oxidants has, therefore, become more essential now. The information on this diverging class of reagents is as scattered as it is rapidly growing and the situation calls for a systematic examination of the published literature. This book attempts a comprehensive and classified compilation of the data that have been accumulated in the last three decades.

References

1. Westheimer FH (1949) Chem Rev 45:419
2. Venkatasubramanian N (1963) J Sci Ind Res 22:397
3. Waters WA (1964) Mechanism of oxidation of organic compounds. Methun and Co Ltd, London
4. Stewart R (1964) Oxidation mechanisms. Benjamin Inc, New York
5. Wiberg KB (1964) Oxidation in organic chemistry. Academic Press Inc, New York
6. Sundaram S, Venkatasubramanian N, Ananatakrishnan SV (1967) J Sci Ind Res 35:518
7. Cainelli G, Cardillo G (1984) Reactivity and structure: concepts in organic chemistry, vol 19. Springer, New York
8. Freeman F (1973) Reviews on the active species in chemical reaction. In: Dayagi S (ed) Chromylchloride oxidations of organic compounds. Freund Publishing Ltd, Israel, pp 39–66
9. Freeman F (1986) Oxidation by oxochromium(VI) compounds; In: Mij's WJ, De Jonge, Cornelis RHI (eds) Organic Syntheses Oxidation by Metallic Compounds, Plenum Press, NY, p 41–118

10. Rocek J (1966) Chemistry of carbonyl compounds. In: Patai S (ed). Interscience, London
11. Bamford CH, Tipper CFH (1971) Comprehensive chemical kinetics, vol 7. Elsevier, New York
12. Corey EJ, Suggs JW (1975) Tetrahedron Lett 2647
13. De Asish (1982) J Sci Ind Res 41:484
14. Piancatelli G, Scettri A, D'Auria M. (1982) Synthesis 245

Chapter 2
Reagent Classification

Abstract The physical characteristics such as the appearance, colour, solubility, melting point, boiling point, of as many as twenty six halochromates besides certain inorganic chromates, dichromates and chromic anhydride are described in this chapter. The laboratory methods for preparing each of these reagents are also discussed.

Keywords Chromate · Halochromates · Alkylammonium chromates · Heterocyclic chromates · Oxidation · Oxidising agent · Chromium oxide

2.1 Dichromates and Chromates

The unique position enjoyed by the acid dichromate and chromic anhydride, at least for a century, as the only good available form of Cr(VI) oxidants is being challenged by the arrival of a host of new complex Cr(VI) reagents from the later half of 1970s. A classified account of such newly emerging reagents on the arena of organic synthesis, their preparative methods and the available physical characteristics are discussed below.

2.1.1 Inorganic dichromates and chromates

The oldest and the most exhaustively evaluated reagent of this category is the acid dichromate which can be easily prepared by dissolving potassium dichromate crystals in concentrated sulphuric acid. Some modified non-aqueous versions of this reagent have also been reported [1]. One of them is prepared *in situ* by dissolving solid $K_2Cr_2O_7$ in DMSO at 50 °C. This solution is stable at room

S. Sundaram and P. S. Raghavan, *Chromium-VI Reagents: Synthetic Applications*, SpringerBriefs in Molecular Science, DOI: 10.1007/978-3-642-20817-1_2, © P. S. Raghavan 2011

temperature but darkens on standing without losing its oxidizing power. The other reagent is prepared in a similar way by dissolving solid $K_2Cr_2O_7$ in polyethyleneglycol-400 (PEG 400) at 50 °C and can be stored at room temperature. Both these reagents have been claimed to be suitable for selective oxidation of allylic and benzylic hydroxyl groups. Also, several solid-supported potassium dichromate reagents have been prepared [2] by deposition of $K_2Cr_2O_7$ on inert solids such as $AlPO_4$, BPO_4 and $AlPO_4$-BPO_4 and their efficiency has been evaluated in the oxidation of cholesterol. Another reagent of this class, chromium peroxidichromate has been well known for its good oxidizing potency [3]. Recent reports reveal that barium and lead chromates are also promising oxidizing systems. It is interesting to note that an aqueous suspension of $BaCrO_4$ is one million times more effective [4] an oxidant for hydrazine compared to the corresponding homogeneous reaction. Sarawadekar et al. [5] have reported on the utility of Ta/$PbCrO_4$ system as an effective oxidant for metallic fuels in pyrotechnic studies.

2.1.2 Alkylammonium dichromates and chromates

These reagents are generally prepared by mixing an aqueous solution of $K_2Cr_2O_7$ or CrO_3 with the corresponding alkylammonium halide. Tetra-n-butylammonium chromate prepared by this method, is a good oxidant for primary and secondary alcohols under non-aqueous and phase tansfer conditions [6]. The corresponding dichromate analogue, bis-tetrabutylammonium dichromate, which is an efficient converter of allylic and benzylic alcohols to the respective aldehydes, may be obtained [7] by mixing a concentrated aqueous solution of $K_2Cr_2O_7$ with twice its molar equivalents of a saturated aqueous tetrabutylammonium bromide solution. Another reagent of this class, bis(benzyltriethylammonium) dichromate has been reported [8] to be a good neutral oxidant for keto alcohols and benzyl halides.

2.1.3 Heterocyclic dichromates and chromates

A wide variety of oxidizing agents of this group with different heterocyclic counter ions have been reported in literature. Each of these has its own merits and demerits and will be discussed in detail below.

(a) Pyridinium dichromate (PDC)

Though this compound has been documented [9–11] for over five decades, a detailed method of preparation, characteristics and utility of this reagent came to be known [12] only in 1979. It was prepared by slow addition of pyridine to a solution of CrO_3 maintained at 30 °C. After the addition is completed, the solution was diluted with acetone and cooled to −20 °C when the orange crystals of PDC appear. It was filtered and dried in air. Alternatively, it can also be prepared [11] *in situ* from ammonium dichromate or an alkali metal dichromate and pyridine hydrochloride. The infrared spectrum of PDC shows the bands characteristic of the dichromate ion at 930, 875, 765 and 730 cm^{-1}. It is sparingly soluble in dichloromethane, acetone and ethanol-free chloroform and not at all soluble in hexane, toluene, ether and ethylacetate. It should be noted that though PDC dissolves in acetonitrile, the solution is not stable. PDC, besides being a good oxidant for simple alcohols, has also been found useful in the oxidation of ω-alkynylalcohols to ω-alkynylaldehydes or ketones [13] and 3,3-disubstituted 1,4-cyclohexadienes to 2,5-cyclohaxadien-1-ones [14].

(b) Polyvinylpyridinium dichromate (PVPDC)

This polymer supported non-acidic reagent can be easily prepared by stirring chromic anhydride into an aqueous suspension of the cross-linked poly(vinylpyridine) resin [15]. It is a recyclable polymeric oxidant which performs very well if used without drying. The presence of water seems to be essential for its functioning, since the dry reagent has a very poor oxidizing capacity. The role of water molecules is to help enhancing the surface area by imbibition. However, a brownish yellow powder form of the reagent could be obtained for storage purposes by drying the wet reagent in vacuo at 50 °C and its activity could be enhanced by soaking it in water prior to use.

(c) Nicotinium dichromate (NDC)

This reagent, also known as 3-carboxypyridinium dichromate, reported [16] in 1986 may be prepared by the slow addition of chromium trioxide to an aqueous solution of 3-carboxypyridine in 1:2 proportions at ice cold conditions. It is a stable, non-hygroscopic and non-photosensitive reagent. It is claimed that this is a good oxidant for both primary and secondary alcohols in binary solvent systems like benzene-pyridine at reflux. 3-Carboxypyridinium dichromate(3-PDC), has also been named as nicotinium dichromate which is misleading [17]; however, it has been rectified in a subsequent report [18].

(d) Quinolinium dichromate (QDC)

QDC is formed as a stable orange solid when an aqueous solution of CrO_3 was added to quinoline at ice cold conditions [19]. It is a mild oxidant and converts primary alcohols exclusively to aldehydes in moderate yields even at room temperature. QDC is sparingly soluble in methylene chloride and chloroform and insoluble in ether, ethylacetate, heptane and toluene. However, it is soluble in polar solvents such as water, DMF and DMSO. Sundar et al. [20] have studied the structural aspects of QDC and have reported that it crystallizes in the monoclinic space group $p2_1/c$, with the unit cell made up of 8 cations and 4 anions.

(e) Tetrakis(pyridine)silver dichromate (TPSDC)

This orange yellow solid could be obtained by mixing a warm aqueous solution of potassium dichromate with a solution of silver nitrate and pyridine [21]. It is reported to be non-photosensitive. It is sparingly soluble in benzene. It must be emphasized here that of all the reported reagents of this family, this is the only oxidant which decomposes in methylene chloride and chloroform.

(f) Imidazolium dichromate (IDC)

Kim and Lhim [22] have reported the synthesis of imidazolium dichromate from imidazole and chromium trioxide. Besides oxidizing alcohols, IDC is also very useful for the deoximation of aldoximes and ketoximes and for converting phenylhydrazones and semicarbazones to the corresponding carbonyl compounds [23].

(g) Benzimidazolium dichromate (BIDC)

When benzimdazole was made to react with chromium trioxide either in water [24] or in acetic acid medium [25], BIDC was obtained. This reagent selectively oxidizes benzylic and allylic alcohols to carbonyl compounds on exposure to microwaves [26].

In addition to those mentioned above, a group of pyridine based oxidants of this class, viz: pyridinium oxychromate, pyridinium oxydichromate and pyridinium sulphodichromate has been reported by Brunelet Thierry et al. [27]. These reagents have been prepared from the appropriate oxide or sulphide compound by treating with CrO_3. The first two among them efficiently convert thiols to disulphides.

2.2 Halochromates

The normal oxidations with dichromates are usually carried out in strongly acidic media. Such severe reaction conditions make it difficult to control the whole process or to isolate the intermediate products. Thus the uncontrollable oxidation or the fragmentation of the products in some cases results in lesser yield of the expected compound. Hence the necessity of oxidizing agents with good efficiency and high selectivity under mild reaction conditions was felt for a long time. The chlorochromate anion was a candidate of choice to suit these requirements. Though the report [28] on the existence of chlorochromate anion dates back to 1833, only in 1952 it was shown [29] that this anion oxidized alcohols less vigorously than the dichromate. Very soon several substituted halochromates started emerging on the scene. All these reagents of the general formula $Q^+CrO_3X^-$, may be prepared from CrO_3, the corresponding base (Q) and the respective hydrohalic acid(HX). A brief account of their preparative methods and characteristics are considered here.

2.2.1 Fluorochromates

(a) Pyridinium fluorochromate (PFC)

Two different methods are available for the preparation of this reagent whose molecular formula is $C_5H_5NH^+CrO_3F^-$. It can be obtained [30] by treating a solution of chromium trioxide in HF with pyridine at ice-cold temperature. Alternatively, it can also be prepared by reacting chromium trioxide with NH_4HF_2 in pyridine [31]. The x-ray diffraction studies on PFC crystals reveal that it has orthorhombic lattice made up of discrete $C_5H_5NH^+$ cations and CrO_3F^- anions. PFC is less acidic than PCC and hence selectively oxidizes secondary hydroxyl groups in the presence of primary hydroxyl groups [32].

(b) Quinolinium fluorochromate (QFC)

This reagent can be synthesized [33, 34] by treating equimolar solutions of quinoline and 40% aqueous HF with a slight excess (1.5 molar) solution of CrO_3. It is quite stable and less acidic than PCC. It is water soluble and also soluble in polar organic solvents. However, it has a very limited solubility in less polar solvents like ether and benzene. QFC is quite suitable for the conversion of alcohols to aldehydes, without further oxidation or any other side reaction. Besides alcohols, thioureas, thioamides, thionoesters and hydrazones are also oxidized by QFC with equal ease [35, 36].

(c) 2,6-Dicarboxypyridinium fluorochromate (DCPFC)

This reagent could be prepared by reacting 2,6-pyridinedicarboxylic acid with chromium trioxide in HF. DCPFC has been shown to be a preferred reagent for the oxidation of alcohols, phenols, hydroquinones [37] and for the oxidative deprotection of oximes, phenylhydrazones, semicarbazones [38] and trimethylsilyl ethers [39].

(d) Benzimidazolium fluorochromate (BIFC)

BIFC could be prepared by treating an equimolar mixture of benzimidazole and chromium oxide with a slight excess of 40% hydrofluoric acid [40]. It is less acidic than PCC—the pH of its 0.01 M aqueous solution being 3.68. It is a good oxidizing agent for any kind of alcohol.

In addition to the above described compounds, many other fluoro chromates have been reported in literature [41, 42] including bipyridinium fluorochromate(BPFC), trimethylammonium fluorochromate [43], tetramethyl ammonium fluorochromate [44], triethylammonium fluoro chromate [45], tripropyl ammonium fluorochromate [46], N-methyl benzyl ammonium fluorochromate [47] and 4-N,N-dimethylaminopyridinium fluorochromate (DMAPFC). In general, these reagents could be prepared by slowly mixing equal quantities of CrO_3 and the respective base in a polythene beaker containing 40% HF. In QFC preparation a slight excess (1.5 mol) of CrO_3 is required. Extra care should be exercised in the preparation of DMAPFC to keep the reaction temperature below 5 °C since at higher temperature a vigorous reaction sets in. The aqueous solution of PFC is acidic in nature (pH of 0.01 M solution is 2.45) and has a fairly good conductance (Molar conductance of 0.001 M solution at 25 °C is 128 mho cm^2 mol^{-1}). Compared with PFC, QFC is less acidic and more soluble in organic solvents.

2.2.2 Chlorochromates

The chlorochromates are the first reported and well characterized reagents among the halochromates. Following the first announcement of pyridinium chlorochromate by Corey [48] in 1975, a dozen of these compounds with counter ions of widely varying shape and structure have been synthesised by different groups of workers and their synthetic and/or kinetic utility investigated. In general, these reagents can be easily prepared by stirring a mixture of CrO_3 and HCl with the base corresponding to the desired counter ion.

(a) Pyridinium chlorochromate (PCC)

This is a stable, non-hygroscopic orange yellow solid. PCC can be prepared by the slow addition of pyridine to an equimolar mixture of CrO_3 and HCl at 0 °C. The pH of an aqueous solution of 0.01 M PCC is 1.75 and thus it is more acidic in nature than PFC. Due to this acidic character of PCC, it might be necessary to buffer the reaction mixture with sodium acetate if any acid labile function in the substrate is to be preserved during the course of the reaction. PCC usually does not affect the simple olefinic bonds and in just 1.5 molar equivalents, it converts primary alcohols exclusively to the respective aldehydes with great efficiency.

(b) 2,2-Bipyridinium chlorochromate (BPCC)

BPCC was first reported [49] by Guziec and Luzzio in 1980. They prepared this reagent by treating 2,2'-bipyridine with chromium trioxide in concentrated HCl. This reagent when preserved over anhydrous $CaCl_2$ and protected from light, retains its efficiency even after storage for three months. Unlike PCC, the internal buffering capacity of the 2,2'-bipyridyl system makes it useful even for acid sensitive substrates. Also, when BPCC is used the chromium byproducts associated with the reactions can be easily removed by filtration through Celite.

(c) 4-N,N-Dimethylaminopyridinium chlorochromate (DMAPCC)

This is a moderately light sensitive reagent and should be protected from light during preparation and storage [50]. DMAPCC is a highly selective oxidant for allylic and benzylic alcohols and the work up is also relatively simple. It is also a mild oxidant suggesting that the introduction of the dimethylamine group increases the basicity of the counter ion and hence reduces the oxidizing power of the anion. The reagent is moderately acidic and in some instances effects the isomerisation of the initially formed Z-unsaturated aldehydes to the corresponding E-isomers. It exchanges with pyridyl carbinol releasing 4-(dimethylamino)pyridine.

(d) Pyrazinium chlorochromate (PzCC)

The non-hygroscopic fine orange needles of this material have a shelf life exceeding three months when stored in dark [51]. It readily exchanges with pyridine forming PCC and free pyrazine. It is insoluble in less polar solvents like methylenechloride, carbon tetrachloride and acetic anhydride. Its oxidizing capacity towards alcohols is relatively low compared to that of PCC.

(e) Naphthyridinium chlorochromate (NCC)

This is a non-hygroscopic, stable orange coloured solid [52]. Its oxidizing potency is far below that of PCC and PzCC. The reagent does not effect any perceptible change on olefinic hydrocarbons such as cyclohexene or allylbenzene.

(f) Tetrabutylammouinm chlorochromate (TBACC)

This is a mild oxidant and can be prepared from CrO_3, HCl and tetrabutyl-ammonium hydrogensulphate [53]. It should be protected from light since it darkens within weeks; however, it does not lose its oxidizing power. At relatively lower molar ratio (3:1) of oxidant to substrate it oxidizes benzylic and allylic alcohols to completion while purely saturated alcohols are not appreciably affected. However, at higher ratio (6:1), saturated alcohols are also significantly oxidized.

(g) Benzyltriethylammonium chlorochromate (BTACC)

This phase transfer catalytic oxidant is reported [54] to have been prepared in three different ways. In one method, an aqueous solution of CrO_3 was added in thin stream to benzyltriethylammonium chloride in water kept at 15 °C. Stirring this mixture for 1 h at 15 °C yielded an orange solid. In the second method, powdered CrO_3 was added in small portions to a well stirred solution of benzyltriethyl-ammonium chloride in methylene chloride solvent. After stirring for an hour, the solvent was removed under reduced pressure while a brown crystalline solid was obtained. This solid material was recrystallised from aqueous acetone to get the reagent identical in all respects to that from the first method. Yet another method has been described for this reagent which proceeds through the formation of PCC. The ultimate yield of the reagent prepared by this method is far greater (95%) than those for the other two methods (73%). In the second method described above, the initial product obtained by the addition of CrO_3 to HCl has been incorrectly assigned [55] a dichromate structure, $(PhCH_2N(C_2H_5)_3)_2Cr_2O_7$. But later it has been proved [54] to be the chlorochromate and not dichromate by quantitatively precipitating the chloride ions from the reagent as AgCl and also preparing the reagent independently from PCC. The melting point of the products obtained from the three methods was in the range 86–90 °C and the products have been well characterized by infra red spectral data. BTACC has been reported to be a good oxidant not only for alcohols but also for aldehydes [56] and carboxylic acids [57].

(h) Guanidinium chlorochromate (GCC)

Though the selectivity of this reagent reported by Santaniello and co-workers [53] is almost the same as that of TBACC, it contaminates the final products because of side reactions and the work up is also difficult.

(i) Methylimidazolium chlorochromate (MICC)

MICC could be prepared [58] by the slow dropwise addition of concentrated hydrochloric acid to an equimolar mixture of N-methylimidazole and chromium trioxide with continuous stirring at 0 °C. This is an efficient oxidizing agent for almost all kinds of alcohols [59]. It is also capable of effecting the oxidative cleavage of oximes.

(j) Trimethylsilyl chlorochromate (TMSCC)

This is a prospective replacement for chromyl chloride in the Etard reaction since it is capable of converting arylmethanes to aromatic aldehydes in good yields. The reagent may be obtained [60] by treating trimethylsilyl chloride with CrO_3.

(k) Quinolinium chlorochromate (QCC)

This is a quite stable yellowish brown solid, prepared by the treatment of chromium trioxide in HCl with quinoline [61]. It is a mild and good selective oxidant for primary alcohols. It can convert alcohols to carbonyl compounds with much better yields in relatively lesser reaction times compared to the conventional methods. Singh et al. [62] has also shown that this is a good reagent for the de-oximation reaction under microwave conditions.

(l) 3-Carboxypyridinium chlorochromate (3CPCC)

This reagent could be prepared by treating 3-carboxypyridine with chromium trioxide in hydrochloric acid. This is a good oxidant under non-aqueous conditions and can be used to convert oximes, phenylhydrozones and semicarbazones to the corresponding carbonyl derivatives in organic solvents [63].

(m) 2,6-Dicarboxypyridinium chlorochromate (DCPCC)

DCPCC can be obtained [64] when pyridine-2,6-dicarboxylic acid is treated with CrO_3 in 6 N HCl. This reagent is soluble in polar solvents, slightly soluble in dichloromethane and chloroform and completely insoluble in tetrachloromethane, benzene and hexane. DCPCC is quite stable at room temperature and retains its oxidation potency for a long period. It is less acidic than 3CPCC—the pH of the 0.01 M aqueous solution of the former is 2.3 while that of the latter is 2.02. This reagent is known to selectively oxidize benzylic alcohols in the presence of other primary and secondary hydroxyl groups [65].

In addition to the above described chlorochromates, γ-picolinium chlorochromate [66], 4-N,N-dimethylaminopyridinium chlorochromate [67], quinaldinium chlorochromate [68], N-methylpiperidinium chlorochromate [69], Prolinium chlorochromate [70], Caffilium chlorochromate [71], and benzyltriethylammonium chlorochromate [72] have also been reported to be good oxidants for alcohols. Trimethylammonium chlorochromate has been suggested [73] to be effective for the conversion of α, β-unsaturated alcohols to the corresponding aldehydes.

2.2.3 Bromochromates

(a) Pyridinium bromochromate (PBC)

This reagent could be prepared as dark brown crystals from CrO_3, pyridine and 47% HBr. It is less acidic (pH of 0.03 M aqueous solution of PBC is 3.35) than PCC or PFC and its aqueous solution has a molar conductivity of 110 mho cm^2 mol^{-1} at 30 °C. PBC functions as a good oxidant as well as a brominating agent [74]. A detailed vibrational assignment of the IR spectrum of the PBC molecule has also been published [75].

(b) Quinolinium bromochromate (QBC)

This oxidant was prepared [76] by a similar procedure as that of QCC with HBr instead of HCl. QBC is a mild and selective oxidizing agent. It was also found to be an effective brominating agent [77].

In addition to the above bromochromates, benzimidazolium bromochromate [78] and triphenylphosphonium bromochromate [27] have also been reported to be good oxidants.

2.2.4 Chromic Anhydride

This is one of the earliest Cr(VI) reagents known in different forms such as Fieser reagent (CrO$_3$-acetic acid), Sarrett reagent (CrO$_3$-pyridine), Conforth reagent (CrO$_3$-pyridine-water), Thiele reagent (CrO$_3$-Ac$_2$O-H$_2$SO$_4$) and Jone's reagent (CrO$_3$-H$_2$SO$_4$) besides its other useful modifications in polar aprotic solvents like DMF, DMSO and HMPT, which have almost replaced the traditional acidic media in organic synthesis for their specific advantages. The utility of these reagents, mainly in organic synthesis, has been well documented in earlier reviews [79, 80]. CrO$_3$ oxidises, besides simple alcohols, C–H bonds present in enolethers [81], epoxides [82], and hydrocarbons [83] to corresponding carbonyl compounds.

In spite of the availability of a wide variety of reagents based on CrO$_3$, the complicated final work up and the maintenance of a strict non-aqueous condition required for some of them necessitated the search for new reagents requiring milder reaction conditions. Following this, several CrO$_3$-amine complexes have come to light such as CrO$_3$-3,5-dimethylpyrazole [84], CrO$_3$-benzotriazole [85] and CrO$_3$-bipyridyl [49]. The CrO$_3$-benzotriazole complex was prepared *in situ* by dissolving benzotriazole in CH$_2$Cl$_2$ containing type 3A molecular sieves and adding an equivalent quantity of CrO$_3$ to this solution. The advantage of using this reagent includes its mild nature, the convenience of reaction at room temperature, the avoidance of special precautions to maintain perfect anhydrous conditions and the augumented selectivity in oxidation of steroidal allylic alcohols. The reddish brown crystalline 2,2'-bipyridine analogue [49] of Colin's reagent was prepared from bipyridine in CH$_2$Cl$_2$ and CrO$_3$ over P$_2$O$_5$. The oxidizing efficiency of this adduct is relatively lower than the corresponding chlorochromate complex (BPCC) and the adduct is a milder reagent since it converts the alcohols exclusively to aldehyde or ketone and no further oxidation to carboxylic acid occurred. Another similar equimolar adduct of CrO$_3$ with chlorotrimethyl silane has been reported by Aizpurua et al. [86] to be a mild oxidant for mercaptans and oximes, but unsuitable for allylic oxidation. Care must be exercised while using this reagent since any distillation involving this reagent results in a violent explosion. However, it can be safely handled in methylene chloride. In addition to the application of CrO$_3$ as a homogeneous oxidant, reports are also available of its use in heterogeneous media with supports such as crown ethers [87] or aluminium silicate [88] or Celite [89].

2.2.5 Miscellaneous

Apart from the aforementioned categories of Cr(VI) oxidants, a few more reagents have been synthesized and evaluated for their individual performance. These

Table 2.1 Physical characteristics of Cr(VI) reagents[a]

S.No	Reagent	Formula	Melting Point	Solvent[b]	Ref.
1	PDC	$(C_5H_5NH^+)_2Cr_2O_7^{2-}$	144–146	H_2O; DMSO; DMF; DMA	[22]
2	PVPDC	$(PVPH)_2Cr_2O_7$	NA	NA	[28]
3	QDC	$(C_9H_7NH^+)_2Cr_2O_7^{2-}$	160–161	H_2O; DMSO; DMF	[32]
4	TPSDC	$(C_5H_5N)_4Ag_2Cr_2O_7$	Decomposes above 113 °C	H_2O; Pyridine; CH_3CN Acetone	[34]
5	PFC	$(C_5H_5NH)CrO_3F$	106–108	H_2O; DMF; Acetone	[43]
6	BPFC	$C_{11}H_9NH^+CrO_3F^-$	114	NA	[55]
7	DMAPFC	$(CH_3)_2NC_5H_4NH^+CrO_3F^-$	184	NA	[55]
8	PCC	$C_6H_5NH^+CrO_3Cl^-$	205(decomp)	NA	[61]
9	PzCC	$C_4H_4N_2HCrO_3Cl$	148–150	H_2O; CH_3CN	[63, 64]
10	NCC	$C_8H_6N_2H^+CrO_3Cl^-$	137–139	H_2O; Pyridine; DMSO; $CHCl_3$; CH_3CN	[63, 64]
11	TBACC	$(C_4H_9)_4N^+CrO_3Cl^-$	NA	THF; $CHCl_3$; EtOAc	[65]
12	BTACC	$C_6H_5CH_2N^+(C_2H_5)_3$ CrO_3Cl^-	86–90	NA	[66]
13	PhCC	$C_{12}H_8N_2H^+CrO_3Cl^-$	Decomposes	NA	[55]
14	PBC	$C_6H_5NH^+CrO_3Br^-$	108	CH_3COOH	[87]
15	Chromyl Chloride	CrO_2Cl_2	115[a]	CCl_4; CS_2	[5]

NA Data not available in the source
[a] Boiling Point
[b] Only the solvents which have been reported to dissolve the reagent yielding a stable solution are listed

include the well characterized chromyl chloride (CrO_2Cl_2) and chromyl acetate, in addition to some Cr(IV) and Cr(V) complexes [90–99]. Chromyl Chloride has also been shown [100] to be useful in the preparation of acetylinic chlorohydrins. Even chromic acid esters of alcohols function as good oxidants. For example, t-butyl chromate prepared by adding CrO_3 in small portions to t-butanol (Caution: flash fire would occur; to avoid this, the accumulation of CrO_3 on the wall of the vessel or on the stirrer should be avoided) at low temperature has been reported to be a novel oxidant [98, 99, 101]. Bis(triphenyl silyl) chromate has been reported [102] to be a new and good methylene oxidant for alkyl ferrocenes. A 1:4 molar adduct of $K_2Cr_2O_7$-chlorotrimethyl silane has been claimed [86] to be a mild and selective oxidant for alcohols and oximes. Thierry et al. [27] have reported the synthesis of triphenylphosphonium oxitrichromate in respectable yield by treating the appropriate oxide with CrO_3. Later, a Cr(V) complex, oxo(salen)chromium(V) where the ligand salen corresponds to N,N′-ethylenebis-salicylidene iminatotrifluoromethane sulphonate, has been claimed [103] to convert alkynes and ylides to the respective carbonyl products. A Cr(IV) complex, oxochrome(IV) tetramethylsilylporphorin, has been shown [104] to be a good oxygen transfer agent.

In Table 2.1 are listed some of the available physical characteristics of all the above chromium oxidants.

References

1. Santaniello E, Ferraboschi P, Sozzani P (1980) Synthesis 646
2. Climent MS, Marinas JM, Sinisterra JV (1987) Tetrahedron 43:3303
3. Wiberg KB (1964) Oxidation in organic chemistry. Academic Press Inc, New York
4. Erwin B, Miguel BA, Ricardo L, Alberto MJG (1985) J Chem Soc Faraday Trans I 81:1113
5. Sarawadekar RG, Menon AR, Banker NS (1983) Thermochim Acta 70:133
6. Cacchi S, La Torra F, Misti D (1979) Synthesis 356
7. Santaniello E, Ferraboschi P (1980) Synth Commun 10:75
8. Miaorong T, Huang X, Chen Z, Hangzhou D X (1984) Ziran Kexucban 11(4):454 CA: 102: 148847b
9. Chevalier P (1942) Rev Cytol Cytophysiol Vegetables 6:221 CA: 45: 9102e
10. Harper AE, Harrison K, Cooke EG (1960) US Patent 2 923 610
11. Coates WM, Corrigan JR (1969) Chem and Ind 1594
12. Corey EJ, Schmidt G (1979) Tetrahedron Lett 399
13. Bierer DE, Koballka GW (1988) Org Prep Proceed Int 20:63
14. Schullz AG, Taveral AG, Harrington RE (1988) Tetrahedron Lett 29(32):3907
15. Frechet Jean MJ, Darling P, Farrall Jean MJ (1981) Org Chem 46:1728
16. Francisco R, Alberto G, Claudo P (1986) Carbohydr Res 149:C1–C4
17. Lopez CM, Gonzalez A, Cossio FP, Palomo C (1985) Synth Commun 15:1197
18. Lopez CM, Gonzalez A, Cossio FP, Palomo C (1987) Tetrahedron 43:3963
19. Balasubramanian K, Prathiba V (1986) Indian J Chem 25B:326
20. Sundar TV, Parthasarathi V, Thamotharan S, Sekar KG (2003) Acta Crystallogr 59:327
21. Firouzabadi H, Sardarian A, Gharibi H (1984) Synth Commun 14:89
22. Kim S, Lhim DC (1986) Bull Chem Soc Jpn 59:3297
23. De SK (2004) Synth Commun 34:2751
24. Meng Q, Feng J, Chin Liu B (1997) J Inorg Chem 13:445

25. Ramiah K, Dubey PK, Ramanathan J, Kumar CR, Grossert JS, Cameron T, Sareda SV (2002) Indian J Chem 41B:2136
26. Meng Q, Feng J, Bian NS, Liu B, Li, CC (1998) 28:1097
27. Brunelet T, Gelband G (1985) J Chem Res Synop 264
28. Peligot A (1933) Ann Chim Phys 52:256
29. Cohen H, Westheimer FH (1952) J Am Chem Soc 74:4387
30. Bhattacharjee MN, Chaudhui MK, Dasgupta HS, Roy N, Khathing D.T (1982) Synthesis 588
31. Chaudhui MK, Chettri SK, Day D, Mandal GC, Paul PC, Kharmawphlang W (1997) J Fluorine Chem 81:211
32. Nonaka T, Kanemoto S, Oshima K, Nozaki H (1984) Bull Chem Soc Jpn 57:2019
33. Murugesan V, Pandurangan A (1992) Indian J Chem 31B:377
34. Chaudhuri MC, Chettri SK, Lyndem S, Paul PC, Srinivas P (1994) Bull Chem Soc Japan 67:1894
35. Bose DS, Arasaiah AV (2000) Synth Commun 30:1153
36. Tajbakhsh M, Mohammedpoor BI, Limohammadi SK, Ramzanian-Lehmali F, Barghamadi M, Shakeri A (2005) Phosphorus, Sulfur, Silicon Relat Elam 180:2587
37. Tajbakhsh M, Hosseinzadeh R, Sadatshahabi M (2005) Synth Commun 35:1243
38. Tajbakhsh M, Hosseinzadeh R, Ramzanian-Lehmali F, Satshahabi M (2005) J Chin Chem Soc 52:1005
39. Tajbakhsh M, Hosseinzadeh R, Ramzanian-Lehmali F, Sadatshahabi M (2005) Phosphorus, sulfur, silicon Relat Elem 180:2279
40. Sivamurugan V, Rajkumar GA, Arabindo B, Murugesan V (2005) Indian J Chem 44B:144
41. Manabendra N. Bhattacharjee, Mihi KC, Himadri SD, Nirmalendu Roy, Darlando TK (1982) Synthesis 588
42. Venimadhavan S, Sundaram S, Venkatasubramanian N (1983) Proceedings of the Reaction Mechanism Groups, 10th Anniversary Meeting, Maynooth, USA 68
43. Sadjadi S AS, Ghammamy S (2006) Indian J Chem 45B:564
44. Hajipour AR, Ruoho AE (2002) Chem Res Synop 547
45. Ghammamy S, Hashemzadeh A, Mazareey M (2005) Russ J Org Chem 42:1752
46. Ghammamy S, Hashemzadeh A (2004) Bull Korean Chem Soc 25:1277
47. Kassaee MZ, Alangi SZS, Ghotbadai HS (2004) Molecules 9:825
48. Corey EJ, Suggs JW (1975) Tetrahedron Lett 2647
49. Frank S Guziec Jr, Frederick A Luzzio (1980) Synthesis 691
50. Frank S Guziec Jr, Frederick A Luzzio (1982) J Org Chem 47:1787
51. Gun LJ, Soo HD, Jong LH (1988) Bull.Korean Chem Soc 9(B):407
52. Harry DB, Roger SM, Jeanne BM, William PW (1983) Heterocycles 20:2029
53. Santanielo E, Milani F, Casati R (1983) Synthesis 749
54. Someswara Rao C, Deshmukh AA, Thakor R, Srinivasan PS (1986) Indian J Chem 25B:324
55. Huang X, Chan C-C (1982) Synthesis 1091
56. Chouhan K, Rao PP, Sharma PK (2006) J Indian Chem Soc 83:191
57. Chouhan K, Sharma P (2004) Indian J Chem 43A:1434
58. Agarwal S, Tiwari HP, Sharma JP (1990) Tetrahedron 46:1963
59. Sanggak K, Heung C (1987) Bull.Korean Chem Soc 8(3):183
60. Jong Gun L, Hee Jong L, Dong Soo H (1987) Bull.Korean Chem Soc 8(5):435
61. Sing J, Kalsi PS, Jawanda GS, Chhabra BR (1986) Chemistry and Industry 751
62. Sing J, Bhandari M, Kaur J, Kad GL (2003) Indian J Chem 42B:405
63. Mohammadpoor-Baltork I, Pouranshirvani Sh (1996) Synth Commun 26:1
64. Hosseinzadeh R, Tajbakhsh M, Niaki MY, Niaki MY (2002) Tetrahedron Lett 43:9413
65. Tajbakhsh M, Hosseinzadeh R, Niak MY (2002) J Chem Res Synop 508
66. Khodaie MM, Salehi P, Goodarzi M (2001) Synth.ommun 31:1253
67. Guziec FSJr, Luzzio FA (1982) J Org Chem 47:1787
68. Degirmenbasi N, Ozgun B (2003) Monatsh Chem 134:1565

69. Tajbaksh M, Ghaemi M, Sarabi S, Ghassemzadesh M, Heravi MM (2000) Monatsh Chem 131:1213
70. Mamaghani M, Shirini F, Parsa F (2002) Russ J Org Chem 38:1113
71. Shirini F, Mohammedpoor-Baltrok I, Hejazi Z, Heravi P (2003) Bull Korean Chem Soc 24:517
72. Someswara Rao C, Deshmukh AA, Thakor MR, Srinivasan PS (1986) Indian J Chem 25B:324
73. Ding Q, Cai K (1988) Youji Huaxue 8(5):457. C.A. 110: 17280w
74. Narayanan N, Balasubramanian TR (1986) Indian J Chem 25B:228
75. Durai S, Ravikumar V, Narayanan N, Balasubramanian TR, Mohan S (1987) Spectrochimica Acta 43A:1191
76. Pandurangan A, Murugesan V, Palanichami MJ (1995) J Indian Chem Soc 72:479
77. Ozgun B, Degrimenbasi N (1996) Synth Commun 29:3601
78. Ozgun B, Degirmenbasi N (1999) Synth Commun 29:763
79. Westheimer FH (1949) Chem Rev 45:419
80. Asish De (1982) J Sci Ind Res 41:484
81. Curci R, Lopez L, Troisi L (1988) Tetrahedron Lett 29(25):3145
82. Hanson JR, Trunesh A (1988) J Chem Soc Perkin Trans 1:2001
83. Linz T, Schaefer HJ (1987) Tetrahedron Lett 28(52):6581
84. Corey EJ, Fleet GWJ (1973) Tetrahedron Lett 4499
85. Parish EJ, Chitrakorn S (1985) Synth Commun 15:393
86. Aizpurua JM, Claudio P (1983) Tetrahedron Lett 24:4367
87. Inaki G, Aizpurua JM, Claudio P (1984) J Chem Res Synop 92
88. Jidong L, Yuanyau W (1987) Synth Commun 17:1717
89. Schmitt SM, Johnston DBR, Christensen BG (1980) J Org Chem 45:1135
90. Krumpolc M, Deboer BB, Rocek J (1978) J Am Chem Soc 100:145
91. Vangalur Srinivasan S, Gould ES (1981) Inorg Chem 20:3176
92. Rajasekhar N, Subramaniam R, Gould ES (1982) Inorg Chem 21:4110; and (1983), 22, 971
93. Bose RN, Gould ES (1985) Inorg Chem 24:2832
94. Fanchiang YT, Bose RN, Gelerinter E, Gould ES (1985) Inorg Chem 24:4679
95. Bose RN, Gould ES (1986) Inorg Chem 25:94
96. Vangalur Srinivasan S (1982) Inorg Chem 21:4328
97. Walba DM, DePuy CH, Garbowski JJ, Bierbaum VM (1984) Organometallics 3:498
98. Roberts DL, Heckman RA, Hege BP, Bellin SA (1968) J Org Chem 33:3566
99. Uma M, Manimekalai A (1985) Indian J Chem 24B:1088
100. Minasyan TT, Grigoryan RT, Badanyan Sh O (1986) Arm Khim Zh 39(12), 763 C.A. 108: 221270p
101. Jacques M (1987) Tetrahedron Lett 28:2133
102. Holecek J, Handlir K, Nadvornik M (1983) J Prakt Chem 325(2): 341; Chem Abstr 99: 175981s
103. Rihter B, Masnovi J (1988) J Chem Soc Chem Commun 35
104. John GT, William KJJr (1985) Isr J Chem 25:148

Chapter 3
Synthetic Strategies

Abstract A critical comparative account of the special properties of the Cr(VI) reagents are discussed in this chapter. These special properties include enhanced potency towards the oxidation of not only alcohols but also other substrates including oximes and aromatic hydrocarbons; their specific synthetic applications for certain regioselective products and modified selective product distribution. A unique and quite unexpected course of reactions such as the hydroxylation at a tertiary carbon atom, observed with some of these reagents are also briefly discussed.

Keywords Synthesis · Regioselective synthesis · Alcohol · Conversion · Reagent potency · Unusual product distribution · Molecular sieve

One has to make a judicious choice of the reagent and the reaction conditions so as to maximize the yield of the intended product. The most sophisticated organic synthesis demands the reagent to possess some specific characteristics such as selectivity, regiospecificity and mildness in reactions. These attributes for a reagent may be ensured either by incorporating modification in its structure or changing the counter ion or by adding some external auxiliary reagents to the reaction medium. In the discussions that follow attention is focused on these aspects of chromium(VI) reagents reported in the literature in the span of last two and a half decades.

3.1 Controlled Conversions

The well established Cr(VI) reagent, viz:chromic acid, when used in the oxidation of alcohols, leads to the formation of carboxylic acids and hardly can one arrest the reaction mid-way. Besides, the poor solubility of potassium and sodium dichromates in water precludes their use without sulphuric acid, therefore, these methods are unsuitable for the oxidation of acid sensitive compounds. Several of the

Table 3.1 Oxidation of benzyl alcohol to benzaldehyde by Cr(VI) reagents

S.No	Reagent	Solvent	Reaction condition[a]	% Yield	Ref.
1	PBC	CHCl$_3$	Reflux; 1.2 eq. 3 h	78	[1]
2	TPSDC	C$_6$H$_6$	Reflux; 2–4 h	80–90	[2]
3	Cl(CH$_3$)$_3$Si–CrO$_3$	CH$_2$Cl$_2$	RT; 45 m	81	[3]
4	TBACC	CHCl$_3$	RT; 3 eq. 4 h	65	[4]
5	NCC	CH$_2$Cl$_2$	1.8 h	50	[5, 6]
6	PzCC	CH$_2$Cl$_2$	1.2 h	50	[5, 6]
7	PCC	CH$_2$Cl$_2$	<0.1 h	50	[5, 6]
8	BPCC	CH$_2$Cl$_2$	2 eq. 2.5 h	79	[7]
9	Bipy-CrO$_3$	Ethylacetate	6 eq. 48 h	90	[7]
10	PFC	CH$_2$Cl$_2$	1.25 eq. 45 m	90	[8]
11	PVPDC	Cyclohexane	1.1 eq. 70 °C; 15 m	75	[9]
12	K$_2$Cr$_2$O$_7$	DMSO	3 h	82	[10]
13	K$_2$Cr$_2$O$_7$	PEG 400	4 h	86	[10]
14	QDC	DMF	1.5 eq. 30 °C	44	[11]
15	QDC	CH$_2$Cl$_2$	1 eq. 4 h	70	[11]
16	3-CPD	CH$_2$Cl$_2$-Pyridine	Reflux	70–90	[12]

[a] The amount of oxidant is given in equivalents relative to that of benzyl alcohol
RT room temperature

reported Cr(VI) reagents encountered in chapter offer various advantages and have been profitably employed in several conversions such as alcohols, oximes, aryl-methanes and alkylhalides to the respective carbonyl compounds, thiols to disul-phides, amines to quinones and aldehydes to esters. Among the reactions reported, a majority involves the conversion of alcohols to aldehydes and for this reason, they are considered first. Benzyl alcohol, which has been chosen as a typical substrate in many investigations, is oxidized efficiently by almost all the reagents in good yield (Table 3.1). These new reagents offer the advantages of easy handling and product separation.

The oxidations of alcohols, other than benzyl alcohol, by the Cr(VI) oxidants are classified in Table 3.2. The reagents are so mild that in all cases a fairly good yield of the respective aldehyde or ketone was obtained and the reaction did not proceed to the carboxylic acid stage. Also, the quantity of oxidant required is low in these oxidations.

Pyridinium chlorochromate has been found to be quite satisfactory for most of the alcohols. However, its acid nature precludes its use for acid-sensitive sub-strates. To circumvent this difficulty, several other organic bases have been tried as counter ions in the place of pyridine. The outcome suggests that the enhanced basicity of the aminic counter part of the potential chlorochromate lowers the oxidizing power of the anion. Thus, for example, bis(tetrabutylammonium) dichromate [27] and 4-N,N-dimethylaminopyridinium chlorochromate [28] turn out to be neutral oxidizing agents. Bipyridinium chlorochromate can be used for oxidations even in the presence of acid-sensitive groups due to its internal buffering and mild character [7]. BPCC offers another advantage in that the excess reagent and the by-product can be removed by a simple dilute acid wash.

Table 3.2 Conversion of alcohols to carbonyl compounds by Cr(VI) reagents

S.No	Substrate	Product	Reagent	Solvent	Conditions	% yield	Ref.
1	Benzhydrol	Benzophenone	PBC	CHCl$_3$	1.2 eq. 3 h; Reflux	83	[1]
2	Benzhydrol	Benzophenone	TBACC	CHCl$_3$	3 eq. 1 h; Reflux	82	[4]
3	Benzhydrol	Benzophenone	PCC	DMSO		100	[13]
4	Benzhydrol	Benzophenone	K$_2$Cr$_2$O$_7$	DMSO	2 h	86	[10]
5	Benzhydrol	Benzophenone	K$_2$Cr$_2$O7	PEG	2 h	86	[10]
6	Benzhydrol	Benzophenone	Cl(CH$_3$)$_3$Si–K$_2$Cr$_2$O$_7$		RT, 20 m	98	[3]
7	Benzhydrol	Benzophenone	QDC	CH$_2$Cl$_2$	1 eq. 4 h	55	[11]
8	Fluorene-9-ol	Fluorene-9-one	PBC	CHCl$_3$	1.2 eq. 4 h; Reflux	70	[1]
9	Benzoin	Benzil	PBC	CHCl$_3$	1.2 eq. 4 h; Reflux	91	[1]
10	Benzoin	Benzil	BTEADC				[14]
11	Benzoin	Benzil	PFC	CH$_2$Cl$_2$	1.5 eq. 1.5 h	92	[8]
12	Benzpinacol	Benzophenone	PBC	CHCl$_3$	1.2 eq.3 h; Reflux	87	[1]
13	4-Methoxybenzyl alcohol	4-Methoxybenzaldehyde	BTEACC	CHCl$_3$	24 h; 25 °C	75	[15]
14	4-Nitrobenzyl alcohol	4-Nitrobenzaldehyde	Cl(CH$_3$)$_3$Si-CrO$_3$	CH$_2$Cl$_2$	RT; 210 m	87	[3]
15	4-Nitrobenzyl alcohol	4-Nitrobenzaldehyde	TBACC	CHCl$_3$	3 eq. 4 h; Reflux	80	[4]
16	4-Chlorobenzyl alcohol	4-Chlorobenzaldehyde	TBACC	CHCl$_3$	3 eq. 8 h; Reflux	78	[4]
17	Cinnamylalcohol	Cinnamaldehyde	TBACC	CHCl$_3$	3 eq. 3 h; Reflux	82	[3]
18	Cyclohexanol	Cyclohexanone	NCC	CH$_2$Cl$_2$	9.2 h	50	[4]
19	Cyclohexanol	Cyclohexanone	FzCC	CH$_2$Cl$_2$	3.2 h	50	[5, 6]
20	Cyclohexanol	Cyclohexanone	PCC	CH$_2$Cl$_2$	0.4 h	50	[5, 6]
21	Cyclohexanol	Cyclohexanone	PFC	CH$_2$Cl$_2$	1.5 eq. 3 h	90	[8]
22	Cyclohexanol	Cyclohexanone	PVPDC	CH	1.1 eq. 70 °C; 24 h	47	[9]
23	Cyclohexanol	Cyclohexanone	PVPDC	CH	1.1 eq. 70 °C; 4 h	69	[9]
24	1-Propanol	Propanal	NCC	CH$_2$Cl$_2$	5.7 h	50	[5, 6]
25	1-Propanol	Propanal	PzCC	CH$_2$Cl$_2$	1.8 h	50	[5, 6]
26	1-Propanol	Propanal	PCC	CH$_2$Cl$_2$	0.1 h	50	[5, 6]
27	2-Cyclohexenol	2-Cyclohexenone	PDC	DMF	1.25 eq.0 °C; 4.5 h	86	[16]
28	Butan-1-ol	Butanal	PFC	CH$_2$Cl$_2$	1.5 eq.2 h	94	[8]

(continued)

Table 3.2 (continued)

S.No	Substrate	Product	Reagent	Solvent	Conditions	% yield	Ref.
29	Butan-1-ol	Butanal	PVPDC	CH	1.1 eq.70 °C; 68 h	81	[9]
30	Butan-1-ol	Butanal	QDC	CH_2Cl_2	1 eq. 4 h; Reflux	69	[11]
31	Pentan-3-ol	Pentan-3-one	PVPDC	CH	1.1 eq.70 °C; 68 h	76	[9]
32	Hexan-1-ol	Hexanal	CrO_3-Bipy	EA	6 eq. 48 h	7	[7]
33	Heptan-1-ol	Heptanal	BPCC	CH_2Cl_2	2.5 eq. 2.5 h	82	[7]
34	Heptan-1-ol	Heptanal	PFC	CH_2Cl_2	1.5 eq. 1 h	84	[8]
35	Octan-1-ol	Octanal	CrO_3	CH_2Cl_2	CE; 0.58 h; RT	40	[17]
36	Octan-1-ol	Octanal	CrO_3	Pyridine		90	[18]
37	Octan-1-ol	Octanal	CrO_3-Bipy	EA	6 eq. 48 h	24	[7]
38	Octan-1-ol	Octanal	BTACC	$CHCl_3$	25 °C; 24 h	30	[15]
39	Decan-1-ol	Decanal	PCC			92	[13]
40	Cinnamylalcohol	Cinnamaldehyde	TPSDC	C_6H_6	24 h; Reflux	80	[2]
41	Cinnamylalcohol	Cinnamaldehyde	PDC	DMF	1.25 eq. 4–5 h; 0 °C	97	[16]
42	Cinnamylalcohol	Cinnamaldehyde	CrO_3Bipy	EA	1 eq. 48 h	81	[7]
43	Cinnamylalcohol	Cinnamaldehyde	PVPDC	CH	1.1 eq. 70 °C; 0.5 h	76	[9]
44	Cinnamylalcohol	Cinnamaldehyde	$K_2Cr_2O_7$	DMSO	3 h	88	[10]
45	Cinnamylalcohol	Cinnamaldehyde	$K_2Cr_2O_7$	PEG	2 h	82	[10]
46	Cinnamylalcohol	Cinnamaldehyde	BTEACC	$CHCl_3$	25 °C; 20 h	45	[15]
47	Cinnamylalcohol	Cinnamaldehyde	QDC	CH_2Cl_2	1 eq; 4 h	70	[11]
48	Cinnamylalcohol	Cinnamaldehyde	QDC	DMF	1.5 eq; 30 °C	52	[11]
49	CH_2=$CHCH_2OH$	CH_2=CHCHO	PCC	CH_2Cl_2	0.1 h	50	[5, 6]
50	CH_2=$CHCH_2OH$	CH_2=CHCHO	NCC	CH_2Cl_2	2.9 h	50	[5, 6]
51	CH_2=$CHCH_2OH$	CH_2=CHCHO	PzCC	CH_2Cl_2	2.9 h	50	[5, 6]
52	$C_6H_5CH_2CH_2OH$	$C_6H_5CH_2CHO$	CrO_3-Bipy	EA	6 eq. 72 h	27	[7]
53	$C_6H_5CH(OH)CH_3$	$C_6H_5COCH_3$	PVPDC	CH	1.1 eq. 70 °C; 5 h	89	[9]
54	$C_6H_5CH(OH)CH_3$	$C_6H_5COCH_3$	BTEACC	$CHCl_3$	20 °C; 20 h	72	[15]
55	$C_6H_5CH_2CH(OH)CH_3$	$C_6H_5CH_2COCH_3$	BTEACC	$CHCl_3$	25 °C; 24 h	15	[15]
56	Nitroalkanol	Nitroalkanal	PCC	CH_2Cl_2	RT; 36 h	61–87	[19]

(continued)

Table 3.2 (continued)

S.No	Substrate	Product	Reagent	Solvent	Conditions	% yield	Ref.
57	Geraneol	Citral	PDC	DMF	1.25 eq. 10 °C; 4.5 h	92	[16]
58	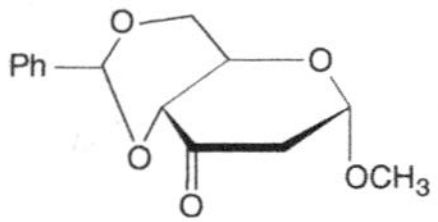 (structure)	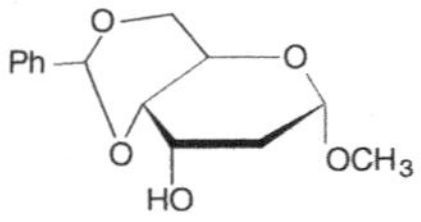 (structure)	PDC	CH₂Cl₂	5 eq. Reflux	70	[20]
59	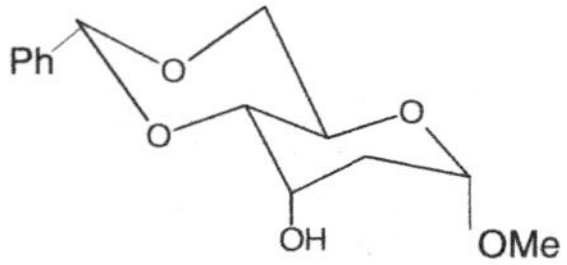 (structure)	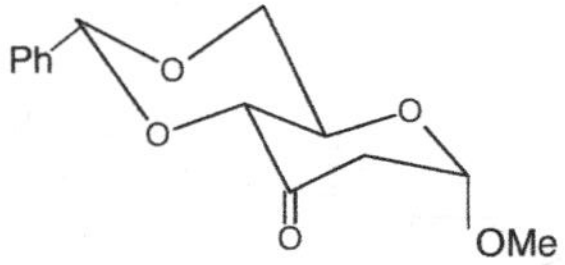 (structure)	PDC	CH₂Cl₂	3A sieve	90	[21]
60	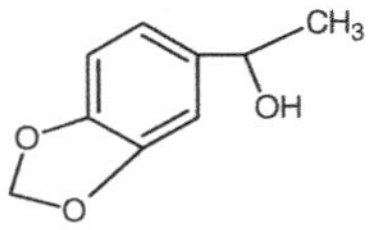 (structure)	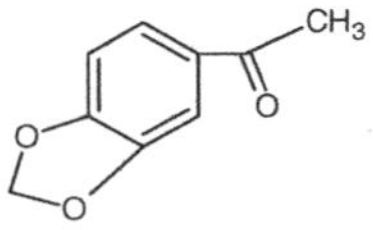 (structure)	BPCC	CH₂Cl₂	3 eq. 3.5 h	95	[7]
61	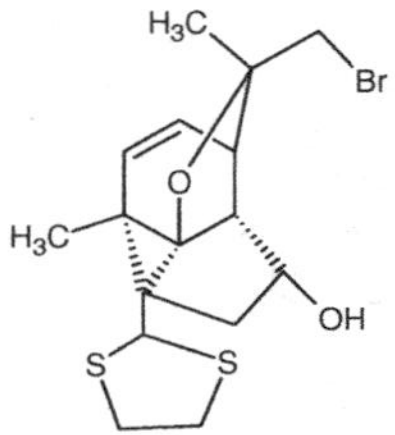 (structure)	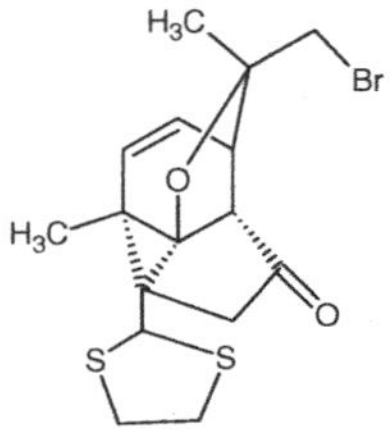 (structure)	PDC	DMF	7 eq. 0 °C; 6 h	95	[16]
62	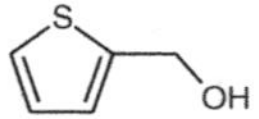 (structure)	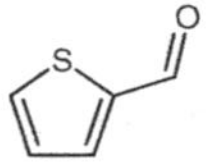 (structure)	BPCC	CH₂Cl₂	3 eq. 3 h	79	[7]

(continued)

Table 3.2 (continued)

S.No	Substrate	Product	Reagent	Solvent	Conditions	% yield	Ref.
63			PCC				[22]
64			CrO$_3$-Py(2:1)	CH$_2$Cl$_2$-DMF(4:1)	1 eq.Ac$_2$O-2 eq. t-BuOH; RT		[23]
65			NDC	C$_6$H$_6$-Py	3 eq. 1 h; Reflux	90	[24]
66	1-Indanol	1-Indanone	Na$_2$Cr$_2$O$_7$	C$_6$H$_6$-H$_2$SO^4		85	[25]
67			PDC	CH$_2$Cl$_2$			[26]

CE crown ether, *CH* cyclohexane, *EA* ethylacetate, *Py* pyridine, *RT* room temperature

Tetrabutylammonium chlorochromate oxidizes allylic and benzylic alcohols completely when refluxed in $CHCl_3$ with a three fold excess of oxidant, while under the same conditions unactivated primary and secondary alcohols are almost unaffected. However, they are incompletely oxidized (50 and 75% respectively) when six equivalents of oxidant were used. The $Cl(CH_3)_3$–Si–CrO_3 adduct is another mild oxidant which cleaves benzaldoxime in modest yield while in the same duration even PCC fails to do so [29] (Scheme 3.1).

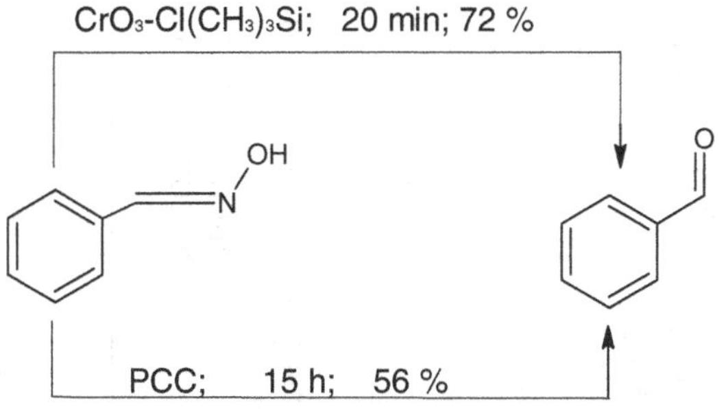

CrO_3-Cl(CH$_3$)$_3$Si; 20 min; 72 %

Scheme 3.1

This reagent, however, fails to effect any change in toluene and, in general, is unsuitable for oxidizing hydrocarbons.

Naphthyridinium and pyrazinium chlorochromates are also good mild oxidants that yield carbonyl compounds from alcohols, without further oxidation to acid [6]. These two reagents are, however, inferior to PCC in potency (entries 19–21, 25–27, and 50–52 in Table 3.2). PDC also performs a controlled conversion of alcohols to carbonyl compounds without involving any further transformation; also, PDC is so mild that it does not affect the acetal and thioacetal units present in the substrate (entries 61 and 66 in Table 3.2). The corresponding nicotinium analogue, nicotinium dichromate (NDC), possesses the characteristics akin to PDC (entry 70 in Table 3.2). PFC has been shown to be efficient in the oxidation of alcohols to aldehydes or ketones. It is claimed to be a good oxygen transfer agent [30], and the final reduction product of the reaction has been shown to be a Cr(IV) species with the formula, $C_5H_5NHCrO_2F$.

Many of these reagents are good methylene oxidants (entries 6–18 in Table 3.3). PFC seems to function more efficiently in polar solvents than in non-polar solvents, since in all the conversions reported PFC gave better yields in acetic acid than in dichloromethane. While anthraquinone and phenanthraquinone are obtained in significant quantities from the oxidation of the respective hydrocarbons by PFC, the response from naphthalene under similar conditions is very poor (only 25% quinine was obtained). This might be due to the reduced methylene character of naphthalene carbons relative to that in the former. QFC also readily oxidizes primary, secondary and allylic alcohols to the corresponding carbonyls, benzoin to benzil, and anthracene and phenanthracene to anthraquinone and 9,10-phenanthraquinone respectively, in dichloromethane solvent. It also converts organic sulfides to

Table 3.3 Conversions involving substrates other than alcohols by Cr(VI) reagents

S.No	Substrate	Product	Reagent	Solvent	Conditions	% Yield	Ref.
1	4-Ntrobenzaldoxime	4-Nitrobenzaldehyde	Cl(CH$_3$)$_3$Si–CrO$_3$	CH$_2$Cl$_2$	15 m; RT	60	[3]
2	4-Ntrobenzaldoxime	4-Nitrobenzaldehyde	PCC/DCCI	CH$_2$Cl$_2$	30 m; T	80	[31]
3	Benzaldoxime	Benzaldehyde	PCC/DCCI	CH$_2$Cl$_2$	30 m; T	85	[31]
4	Pyridine-4-aldoxime	Pyridine-4-aldehyde	PCC/DCCI	CH$_2$Cl$_2$	30 m; T	85	[31]
5	R$_2$C=N–OH	R$_2$C=O	PCC/H$_2$O$_2$		0–10 m	85	[3]
6	R–C$_6$H$_4$–CH$_3$ (R=H; 2–CH$_3$; 4–CH$_3$; 2–Br; 3–OCH$_3$)	R–C$_6$H$_4$CHO	TMSCC	CCl$_4$		44–71	[32]
7	FcCH$_2$R (R=CH$_3$; C$_6$H$_4$; Fc Fc=ferrocenyl)	FcCOR	BTPSC			18–54	[33]
8	FcR=CHR (R=CH$_3$; C$_6$H$_4$; Fc Fc=ferrocenyl)	FcCOR	BTPSC				[33]
9	5,6-Dihydropyran	5,6-Dihydropyranone	PCC			60–85	[34]
10	Isochroman	Isochromanone	PCC			80	[34]
11	Indan	1-Indanone	Chromic acid		6 eq.		[35]
12	Tetralin	1-Tetralone	Chromic acid		6 eq.		[35]
13	Anthracene	9,10-Anthraquinone	PFC	CH$_2$Cl$_2$	2.5 eq. 4 h	68	[8]
14	Anthracene	9,10-Anthraquinone	PFC	CH$_3$COOH	2.5 eq. 90 m	98	[8]
15	Phenanthrene	9,10-Phenanthraquinone	PFC	CH$_2$Cl$_2$	2.5 eq. 5 h	52	[8]

(continued)

Table 3.3 (continued)

S.No	Substrate	Product	Reagent	Solvent	Conditions	% Yield	Ref.
16	Phenanthrene	9,10-Phenanthraquinone	PFC	CH₃COOH	2.5 eq. 3 h	72	[8]
17	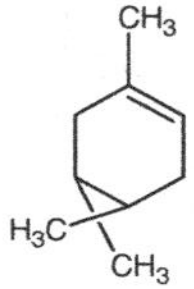 (structure)	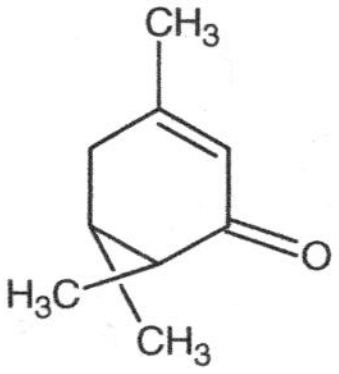 (structure)	PFC		Acid medium	45	[36]
18	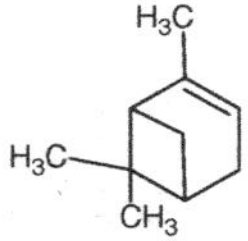 (structure)	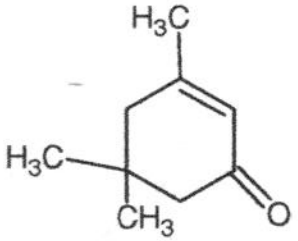 (structure)	PFC			35	[36]
19	Benzyl halide	Benzaldehyde	BTEADC			11	[14]
20	Benzyl halide	Benzaldehyde	K₂Cr₂O₇	HMPT	BTAC	100	[37]
21	Butyl chloride	Butanal + Butan-1-ol	K₂Cr₂O₇	HMPT	BTAC	18	[37]
22	Allyl chloride	Acrolein + Allyl alcohol	K₂Cr₂O₇	HMPT	BTAC	80	[37]
23	Octyl bromide	Octanal + Octan-1-ol	K₂Cr₂O₇	HMPT	BTAC	2	[37]
24	Octyl iodide	Octanal + Octan-1-ol	K₂Cr₂O₇	HMPT	BTAC	20	[37]
25	RCH₂Br (R=butyl; pentyl; hexyl;heptyl)	RCHO + RCH₂OH	K₂Cr₂O₇	DMF	Crown ether; 150 °C; 2 h	28–51	[38]
26	RSH	R–S–S–R	TBACC	CHCl₃	0.5 eq. RT	82–87	[4]
27	C₆H₄SH	C₆H₄–S–S–C₆H₄	TMSCC	CHCl₃	13 °C; 15 m;	93	[3]
28	RSH	R–S–S–R	PDC	CHCl₃			[39]
29	R–C₆H₄NH₂	Substituted ortho or para Benzoquinone	CrO₂Cl₂	CCl₄	0–5 °C; 1 h	14–24	[40]

(continued)

Table 3.3 (continued)

S.No	Substrate	Product	Reagent	Solvent	Conditions	% Yield	Ref.
30	2,3-Dimethylaniline	2,3-Dimethylbenzoquinone	$Na_2Cr_2O_7$–H_2SO_4		Silica gel; 60–80 °C; 1 h	94	[41]
31	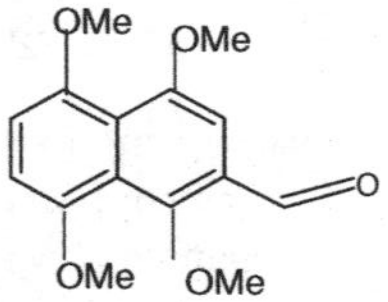	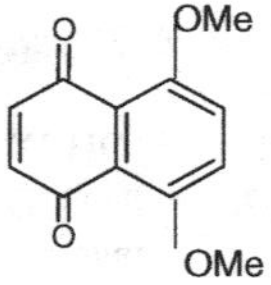	Jone's Reagent	Acetone	RT	89	[42]
32	Ph_3P	$Ph_3P = O$	PFC	CH_3CN			[43]
33	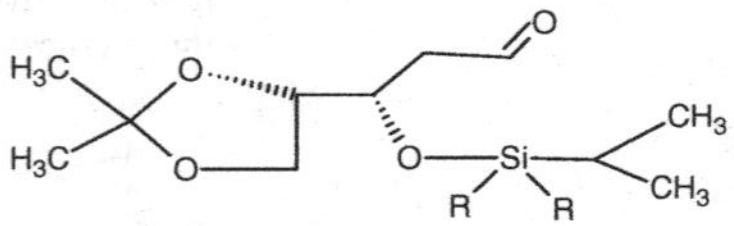	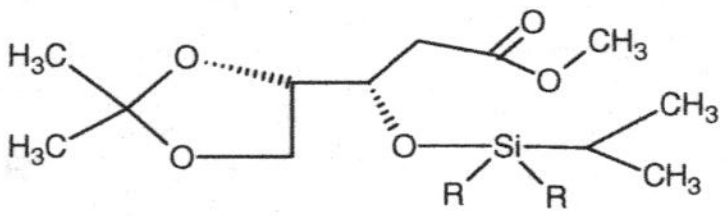	PDC/CH_3OH	DMF	6 eq. RT	64	[44]
34	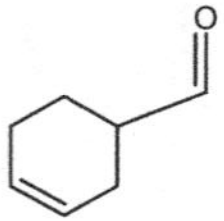	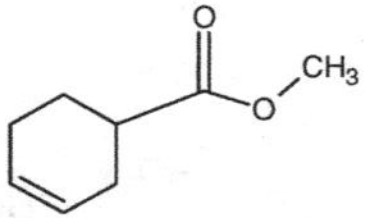	PDC/CH_3OH	DMF	6 eq. RT	80	[44]
35	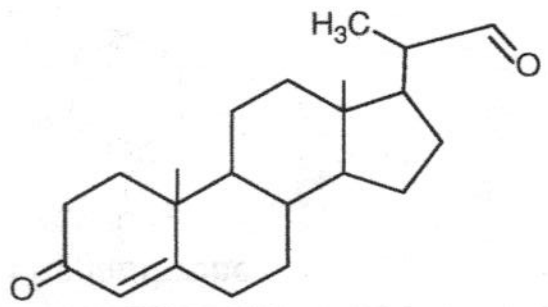	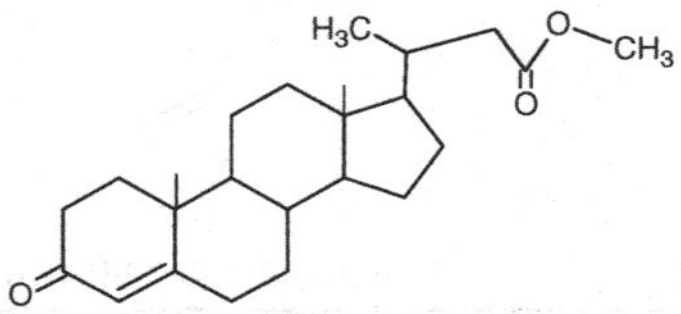	PDC/CH_3OH	DMF	6 eq. RT	60	[44]
36	Cycloalkylboranes	Cycloalkylketones	Chromic acid				[45]

(continued)

Table 3.3 (continued)

S.No	Substrate	Product	Reagent	Solvent	Conditions	% Yield	Ref.
37	Oxiranes	C–C cleavage products	Chromic acid				[46]
38	Coal	Carboxylic acids	Chromic acid				[47]
39			Chromic acid			30	[48]
40	$R_2C = CR'CHPh_2$, R = Ph; R' = H, CH_3-		CrO_3-Ac_2O		H_2SO_4 medium		[49]
41	R-Substituted, p-Quinols (R = H,Cl,CH_3, OCH_3)	R-Substituted, p-benzoquinone	3-CPD	Pyridine – Benzene	Reflux	70–100	[12]

Table 3.4 Product distribution in the oxidation of alkylhalides with $K_2Cr_2O_7$ in HMPT[a]

R–CH$_2$–X		Product percentage		
R =	X =	R–CHO	R–CH$_2$–OH	R–CH$_2$–X[b]
C$_6$H$_4$–	Cl, I	100		
n-Butyl	Cl	18	80	2
Allyl	Cl	10	70	20
Octyl	Br	2	76	22
Octyl	I	20	40	40

[a] Data collected from reference 217
[b] Unreacted substrate

sulfoxides at room temperature. In all its oxidations the final reduction product from QFC has been ascertained to be $C_9H_7NH(CrO_3F)$, a Cr(IV) species.

In an effort to eliminate acid medium, the oxidation with dichromate has been attempted by various workers in several dipolar aprotic solvents like DMSO, HMPT, DMH and PEG. Switching over to such solvents confers a specificity on the dichromate ion. Thus $K_2Cr_2O_7$ in either DMSO or PEG-400 brings about a specific oxidation of the allylic and benzylic alcohols; the saturated alcohols are not at all affected. Similarly, $K_2Cr_2O_7$ in HMPT or DMF is an excellent oxidant for benzylic halides but with aliphatic and saturated halides most of the unreacted substrate and partial oxidation product (alcohol) were recovered from the reaction mixture (see Table 3.4). Bis (tetra-n-butylammonium) dichromate is also specific for allylic and benzylic alcohols while unactivated alcohols resist attack by this reagent [1].

Chromyl chloride has been well known [50, 51] since the discovery of Etard's reaction as one of the best reagents to yield aldehyde from substituted aromatic hydrocarbons. Nalliah and Strickson have reported [52] that the Etard adducts of aromatic primary amines with CrO_2Cl_2 in CCl_4 or $CHCl_3$ on hydrolysis result in the formation of a mixture of azobenzenes, 1,4-benzoquinones, 1,4-benzoquinone anils and aniline-1,4-benzoquinone. The composition of the mixture depends on the nature of the reactant. However, the parent molecule, aniline, leads to an intractable polymeric material. A subsequent report [53] characterises these products as 1:1(**1**) or 1:2(**2**) stoichiometric adducts containing imino or diimino ligands with a giant structure formed by chloro or hydroxyl bridges. These adducts are insoluble in non-polar solvents while they react irreversibly with polar solvents.

The adducts **1** and **2** are coloured, the colour varying from deep brown to greenish blue depending on the nature of the aromatic amine. The tracer studies with O^{18} on these adducts showed that Cr(IV) is the central metal in them and that the oxygen atom of the quinine anil and one of the oxygen atoms of quinine are derived from water during hydrolysis.

Apart from their ability to convert alcohol to aldehyde or ketone, some reagents function unusually in different environments. For example, PDC when used in conjunction with methanol in DMF leads to a convenient one-step conversion of an alcohol or aldehyde to the methyl ester at ambient conditions (entries 33–35 in Table 3.3). A similar one-step conversion of alcohols to their

(1)   (2)

Table 3.5 Bromination by pyridinium bromochromate[a]

Substrate	Product	Solvent	Conditions	% Yield
Anisole	4-bromoanisole	Acetic acid	90–100 °C; 1 h	89
Anisole	4-bromoanisole	Chloroform	90–100 °C; 3 h	63
Acetanilide	4-bromoacetanilide	Acetic acid	90–100 °C; 0.5 h	87
Acetanilide	4-bromoacetanilide	Chloroform	90–100 °C; 2 h	85
Acetophenone	ω-bromoacetophenone	Acetic acid	90–100 °C; 4 h	40
Anthracene	9,10-dibromoanthracene	Acetic acid	90–100 °C; 0.5 h	84
Fluorene	9-bromofluorene	Acetic acid	90–100 °C; 3 h	73

[a] Data collected from reference 1

t-butyl ester is possible (entry 69, Table 3.2) by reacting the alcohol with four equivalents of 2:1 CrO_3-pyridine in 4:1 CH_2Cl_2-DMF at room temperature, in presence of t-butanol and acetic anhydride. The reaction possibly proceeds via the hemiacetal with t-butanol. That such an ester formation does not occur with unreactive aldehydes like piperonal supports this suggestion. PBC plays a dual role as an oxidizing and brominating agent. Its role as a brominating or an oxidizing agent depends upon the nature of the substrate and the polarity of the reaction medium. More specifically, for substrates with high electron density, it is a brominating agent. Thus anisole, acetanilide, acetophenone and the hydrocarbons anthracene and fluorene are reported to give the brominated product in moderate to good yields (Table 3.5). The bromination is more facile in polar solvent (acetic acid) than in less polar (chloroform) media. Grigor'ev et al. [54] have compared the oxidizing potency of ammonium and alkali metal dichromates in the iodination of aromatic compounds. Their results indicate that the counter cation has an appreciable influence on the oxidizing ability of the dichromate ion and it followed the order

$$K_2Cr_2O_7 < Na_2Cr_2O_7 < Li_2Cr_2O_7 < (NH_4)_2Cr_2O_7.$$

3.2 Regioselective Synthesis

The success of a newly discovered reagent as an acceptable or preferred synthetic oxidant mainly relies on its efficacy to select a single target of attack among more than one available sites. Regioselectivity is rather an imminent property expected of a reagent in the synthetic design of natural products and products of biological interest. Such selective and specific transformations are generally difficult to accomplish, especially in high yields. For example, manganese dioxide is generally used to bring about the selective oxidation of allylic and benzylic alcohols [55–57]. However, undesirable side reactions, abnormal reaction times and the complications associated with the oxidation of hindered substrates have restricted it from being popular. Dichlorodicyanoquinone (DDQ) and some chromate oxidizing reagents [28, 58, 59] have also been reported to selectively oxidize benzylic and allylic alcohols. But, these reactions also suffer from extended reaction times and only moderate selectivity.

Consequent to the realization of the acidity of the Corey's reagent (PCC) its combination with different organic bases have been experimented within various oxidations. Such a combination alleviates or minimizes the harmful acidic effect of PCC and makes it suitable even for substrates containing acid-sensitive functions. Organic bases such as pyridine [60], pyrazole [61, 62], 2.2′-bipyridine, pyrazine, pyridazine, 2,4,6-tri phenylpyridine, S-triazine [63] and benzotriazle [64] when used in combination with PCC, confer regioselectivity. PCC, with any of these amines at 2% level is a mild and selective oxidant for allylic and benzylic hydroxyl groups. According to Parrish and Schroepfer [60], the selectivity of the reagent and hence the reaction product vary depending upon the nature of the base employed (Scheme 3.2).

The appropriate combination of the PCC-amine system not only reduces the time involved in the conversions, driving the reactions to near completion at low temperature(2–3 °C), but also is very specific in their attack (entries 8, 9, 15, 17 and 20 in Table 3.6). A similar selective affinity towards allylic [65] and benzylic hydroxyl groups as against the complete failure in the attempted oxidation of saturated alcohols has been reported for $K_2Cr_2O_7$ in DMSO or PEG [10]. PCC has also been recommended [66] for selective oxidation of hydroxyl group in the 3β position of sterols in preference to that in the 3α position.

In the synthesis of butenolides and benzofuranones, PCC has been employed [75] for the conversion of reactive methylene group to carbonyl group, the yields being 35 and 85% respectively, for the two systems. A direct conversion of aldehydes to carbamoyl azides [76], a regioselective conversion [77] of 1-methyl-cyclooct-4-en-1-ol to 1-methyl-oxabicyclo [4.2.1] nonan-5-one, manufacture [78] of α-chloroacetophenone from 2-chloro-1-phenyl ethanol and preparation [79] of cyclopentanones from cyclo-pentadienes have also been successfully demonstrated to be the synthetic utility of PCC.

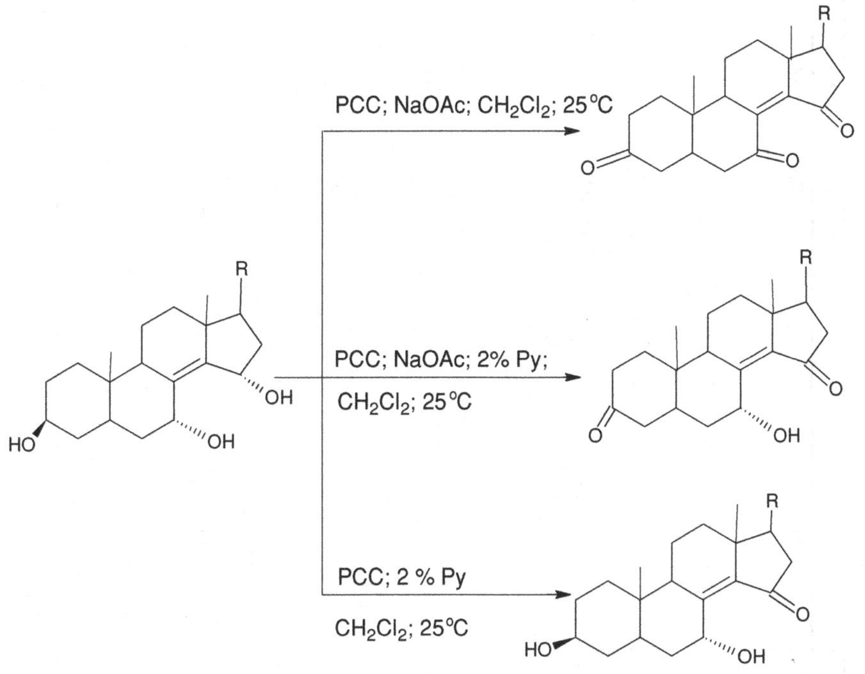

Scheme 3.2 Regioselective oxidation of 5α-cholest-8(14)-ene-3β, 7α, 15α-triol with PCC

Roberto et al. [80] have reported that pyridinium dichromate containing 1–5 mmoles of iodine effects a stereospecific conversion of olefine to the corresponding iodohydrin or epoxide depending upon the reaction conditions. They have attributed the regiospecificity of these reactions to the formation of an iodonium ion in the first step, which might undergo a preferential nucleophilic attack by the dichromate ion on the tertiary carbon atom rather than on the secondary one.

A similar regioselective α-iodohydrin formation with terminal and internal olefinic esters effected by Ag_2CrO_4 and I_2 in proper ratio has been reported more recently [81]. Chromic anhydride has also been reported to yield stereospecific products in an unusual solvent mixture (entry 12 in Table 3.6) and under phase transfer conditions (entry 13 in Table 3.6).

3.3 Enhancement of Reagent Potency

The relatively more recent Cr(VI) reagents discussed earlier have been claimed to be mild in potency and convenient to handle; but, at least, some would require a prolonged period for achieving complete conversions. To reduce the duration of such reactions and to improve the potency of the oxidant, the addition of a variety

Table 3.6 Specific products in the oxidations by Cr(VI) reagents

S.No	Substrate	Product(s)	% Yield	Reagent	Solvent	Ref.
1	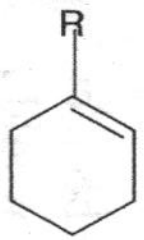	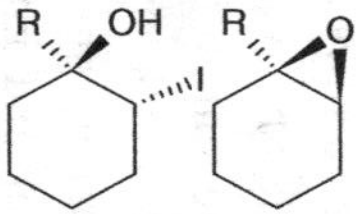 $R = C_5H_{11}$-	50–60	PDC/I_2	CH_2Cl_2	[53]
2	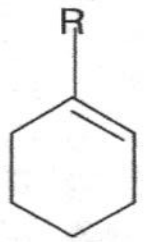	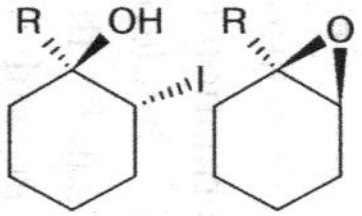 $R = C_{10}H_{21}$-	50–60	PDC/I_2	CH_2Cl_2	[53]
3	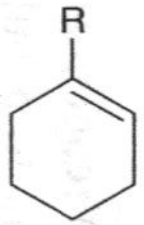	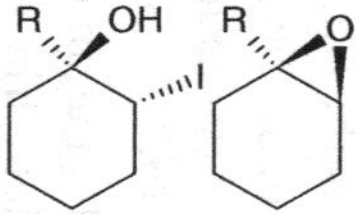 $R = C_6H_5$-	50–60	PDC/I_2	CH_2Cl_2	[53]
4	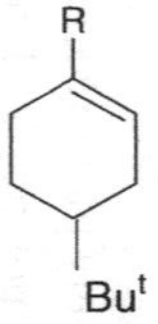	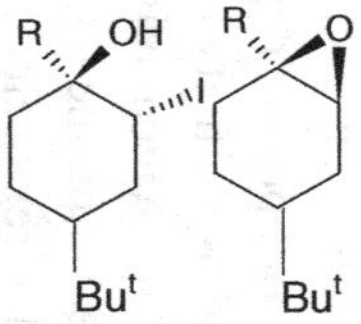 $R = C_9H_{19}$-	50–60	PDC/I_2	CH_2Cl_2	[53]

(continued)

Table 3.6 (continued)

S.No	Substrate	Product(s)	% Yield	Reagent	Solvent	Ref.
5	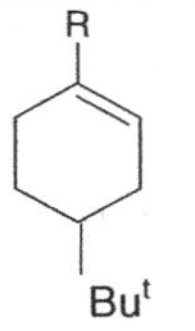	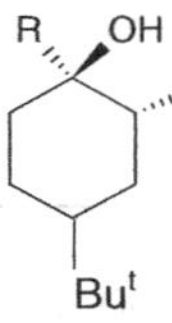 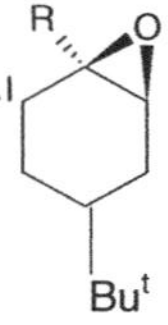	10	PDC/I$_2$	CH$_2$Cl$_2$	[53]
6	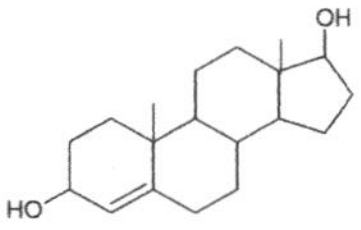	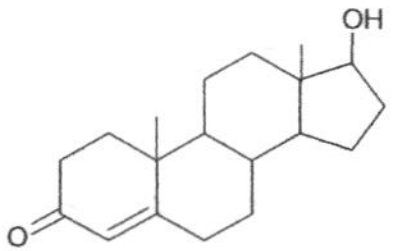	88	PCC-Pyrazole	CH$_2$Cl$_2$	[67, 68]
7	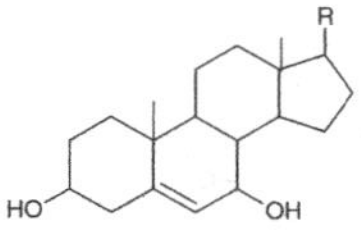	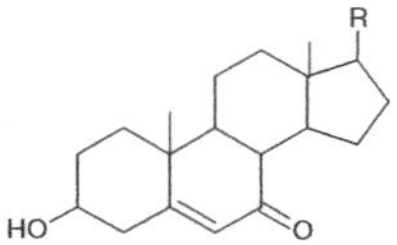	93	PCC-Pyrazol;e	CH$_2$Cl$_2$	[67, 68]
8	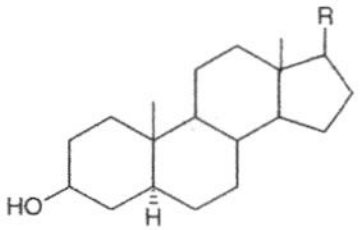	Poor reaction[a]		PCC-Pyrazole	CH$_2$Cl$_2$	[67, 68]
9	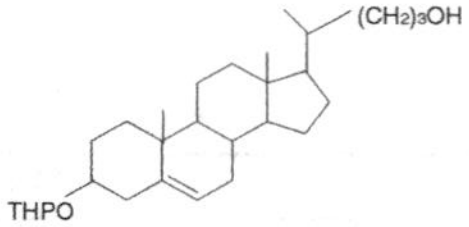	Poor reaction[a]		PCC-Pyrazole	CH$_2$Cl$_2$	[67, 68]

(continued)

Table 3.6 (continued)

S.No	Substrate	Product(s)	% Yield	Reagent	Solvent	Ref.
10	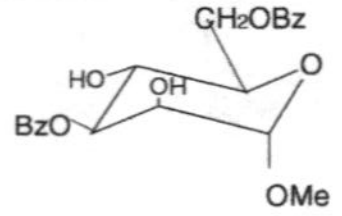	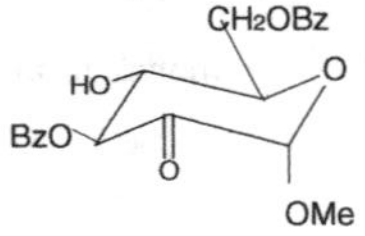		PCC		[69]
11	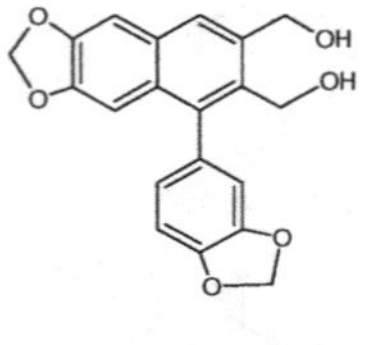	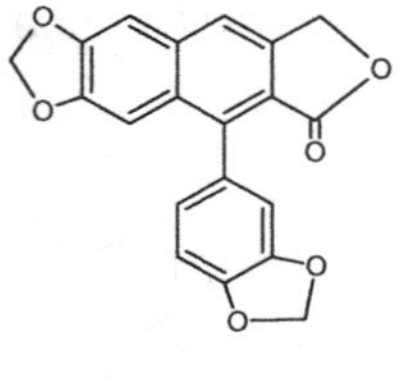 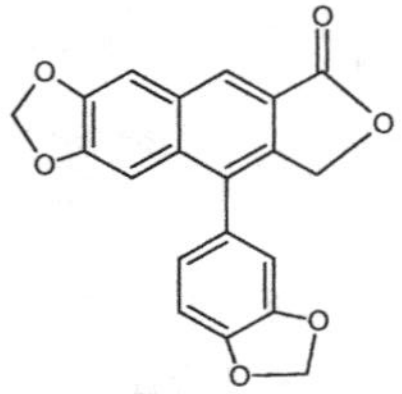	67 33	PCC		[22]
12	$CH_3(CH_2)_{13}CH_2CH = CHCOOCH_3$	$CH_3(CH_2)_{13}COCH=CHCOOCH_3$	75	CrO_3	C_6H_6- CH_3COOH- $(CH_3CO)_2O$	[70]

(continued)

Table 3.6 (continued)

S.No	Substrate	Product(s)	% Yield	Reagent	Solvent	Ref.
13	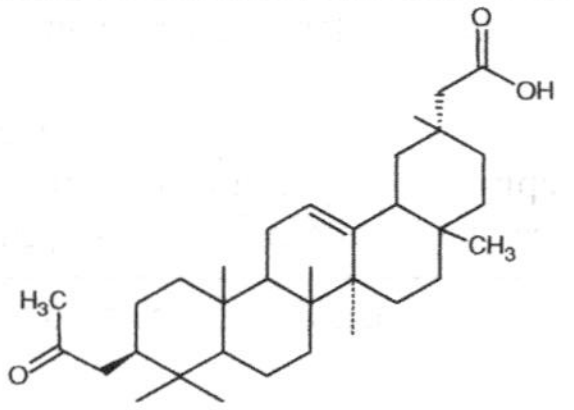	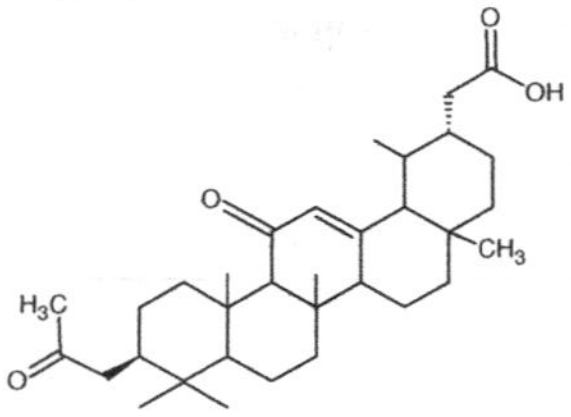	80	CrO$_3$–CTAB	CCl$_4$	[71]
14	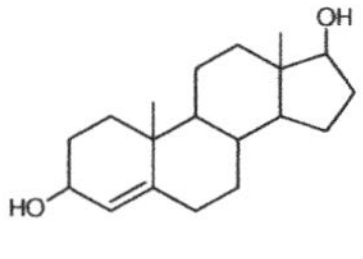	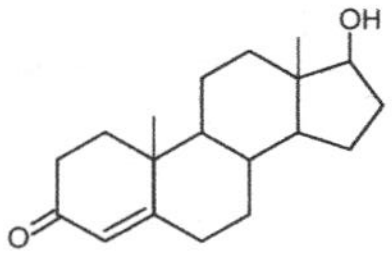	92	PCC/BT^c	CH$_2$Cl$_2$	[72]
15	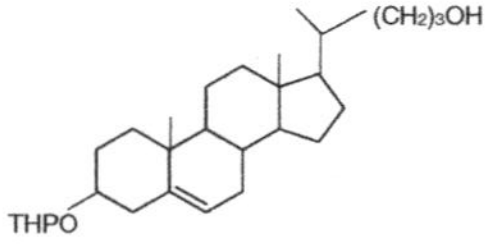	Poor reaction^a		PCC/BT	CH$_2$Cl$_2$	[72]
16	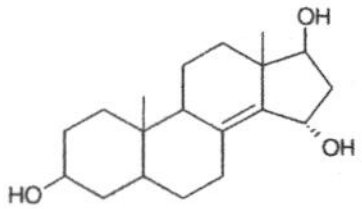	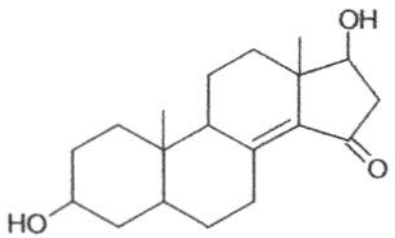	87	PCC-Py	CH$_2$Cl$_2$	[73]

(continued)

Table 3.6 (continued)

S.No	Substrate	Product(s)	% Yield	Reagent	Solvent	Ref.
17		Poor reaction		PCC-Py	CH_2Cl_2	[73]
18			88 91 90 55(36)[e]	PCC-2,2'-Bipy[d] PCC-Py PCC-Pyradizine PCC-2,4,6-Triphenylpyridine	CH_2Cl_2 CH_2Cl_2 CH_2Cl_2 CH_2Cl_2	[30]
19		$R = C_8H_{17}-$ [f]	59(32)[e] 60(32)[e] 39(51)[e] 20(72)[e] 11(84)[e]	PCC-Bipy PCC-Pyrazine PCC-Pyradizine PCC/S-triazine PCC/ 2,4,6-Triphenylpyridine	CH_2Cl_2 CH_2Cl_2	[30]
20		Poor reaction			CH_2Cl_2	[30]
21	Cholesteryl benzoate	7-Oxocholesteryl benzoate	87	PCC	C_6H_6	[74]

[a] About 90–95% of the reactant recovered at the end of the reaction
[b] Cetyltrimethylammonium bromide
[c] Benzotriazole
[d] Bipyridyl
[e] Percentage of the reactant recovered at the end of the reaction is given in the parenthesis
[f] About 90% of the substrate recovered at the end of the reaction with all the five amine-PCC reagents mentioned in entry 19

of auxiallary reagents in catalytic quantities has been experimented with. For example, Corey and Schmidt have reported [16] that the addition of a small quantity of pyridinium trifluoroacetate(PTFA) to PDC reduces the quantity of oxidant required as also the reaction time. According to a later report [82], anhydrous acetic acid is a better promoter than PTFA for the same reaction. Further, a comparative study on the promoter efficiency of PTFA, anhydrous acetic acid and some molecular sieves in the oxidation of carbohydrates by PDC has revealed that a judicious combination of anhydrous acetic acid and 3A molecular sieve could be dramatic in its action (Scheme 3.3). Benzhydrol was converted to benzophenone in just two min in 91% yield when the combination of 100 μL

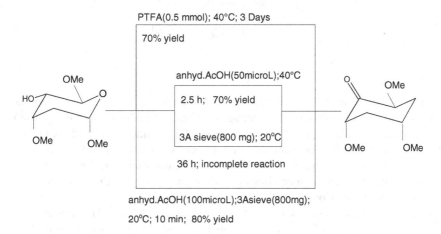

Scheme 3.3 All reactions carried out with 1.5 equivalence of PDC in CH_2Cl_2

anhydrous acetic acid and 800 mg of freshly activated 3A molecular sieve was employed as the promoter.

Herscovici et al. [64] have recommended such molecular sieves as good potency improvers for PCC.

In the synthesis of some ketonucleosides, PCC by itself, surprisingly did not work for the substrate 7-(3-bromo-3,4,6-trideoxy-L-erythrohex-3-eno pyranosyl) theophylline (**3**) and the addition of molecular sieves to the reaction mixture led to

R= theophylline-7-yl

(**3**)

the immediate precipitation of the reduced chromium species attended by the formation of the desired product.

The potency of PCC is also significantly augmented when it is used in conjunction with H_2O_2. Thus, the reagent system containing PCC with 30% H_2O_2 cleaved several ketoximes within 10 min at 0–10 °C with 60–85% yield while PCC alone took 19–94 min depending upon the nature of the ketoximes [83]. It has been suggested that the active deoximation species in this reagent system could be the pyridinium oxodiperoxy chlorochromate(**4**) generated in situ from PCC and H_2O_2.

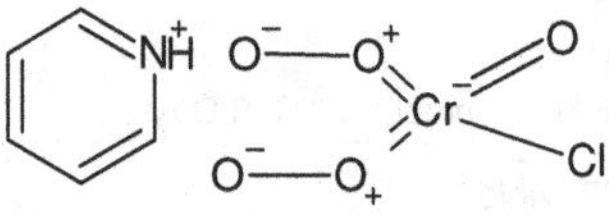

$$\text{(structure 4)}$$

(4)

The presence of fluoride ion has also been reported [84] to facilitate the cleavage of Si–O bonds by Jone'sreagent. Thus the addition of KF to Jone's reagent enabled it to oxidize t-butyldimethylsilyl and trimethylsilyl ethers in 98% yield just in 1 h at 0 °C in acetone to the respective ketones or carboxylic acids. However, t-butyl diphenyl silyl ether resists oxidation under similar conditions.

In contrast to such augmentations of oxidation potency of Cr(VI) reagents by auxiallary reagents, Corey et al. [85] have made an unusual observation that the addition of the chromate ester of 2.4-dimethylpentane-2,4-diol(**5**) efficiently catalysed the oxidation of secondary alcohols to ketones by peroxyacetic acid.

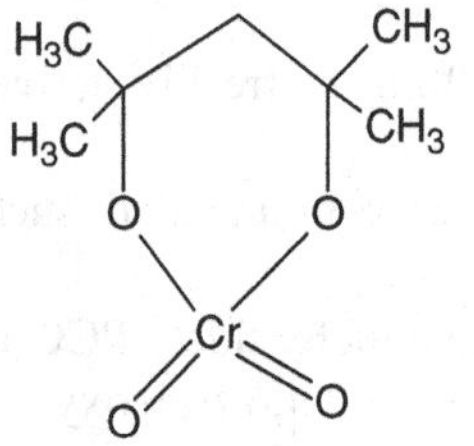

$$\text{(structure 5)}$$

(5)

3.4 Unusual Product Distribution

It is interesting to learn that a sizeable percentage of certain reactions with these reagents are diverted through an unusual path leading to the formation of other novel, unexpected products. For example, when α-ionone was oxidized with t-butyl chromate, about 22% of tertiary carbon–hydroxylated product was formed besides the usual allylic oxidation [86]. On the other hand, when vasicine was subjected to reaction with PFC, the major product (45%) obtained involved methylene oxidation, leaving the hydroxyl group intact and, the expected diketone was not al all formed [87]. Paquette et al. [88] have observed an unprecedented oxidative process in the reaction of seco-dodcahedrene (6). When 6 waas oxidized by $Na_2Cr_2O_7$ in acetic acid-acetic anhydride medium, an unusually regiospecific reaction occurred and a C_2-symmetric diepoxide was formed (7) while oxidation of (6) with m-chloroperbenzoic acid (mcPBA) yielded the simple mono epoxide, (8) (entry 4; Table 3.7).

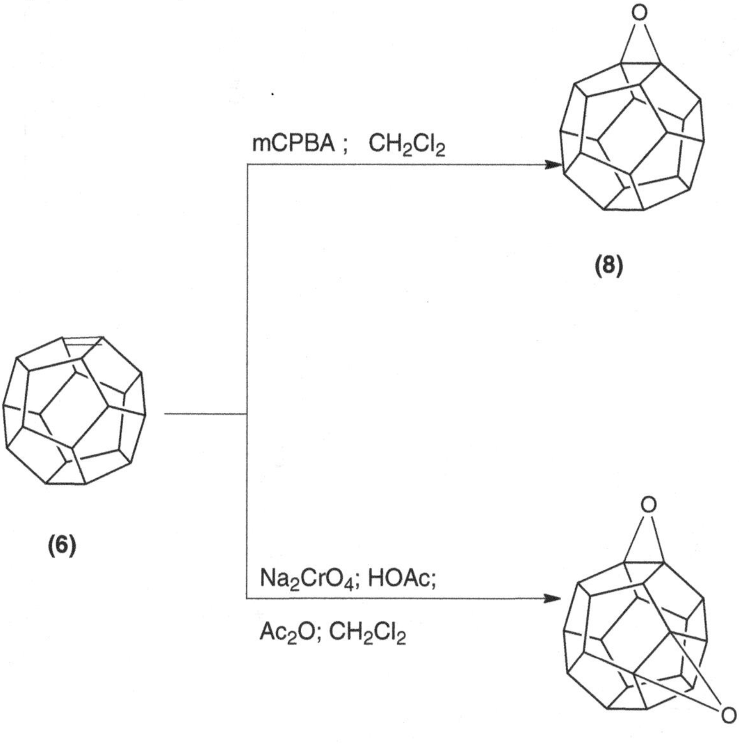

Similarly, when 1,3,5-cycloheptatriene (9) was treated with singlet oxygen followed by the treatment with CrO_3 in acetone, it yielded [89] 3-hydroxy tropone (11) via the intermediate compound (10). Though the transformation of (10–11)

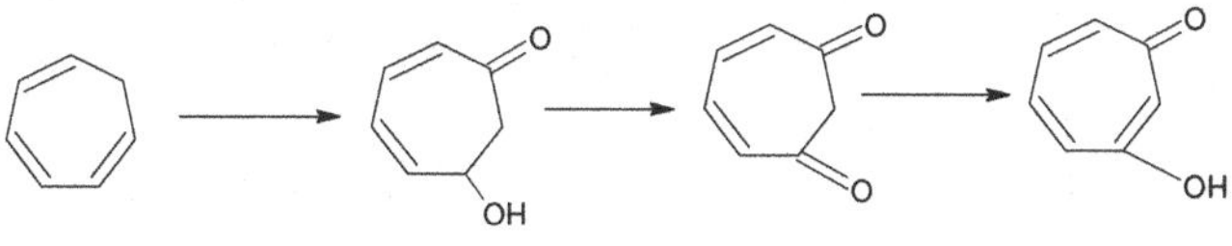

Scheme 3.4 Conversion of 1,3,5-cycloheptatriene to 3-hydroxy tropone

Table 3.7 Unexpected Product(s) in the Oxidation reactions with Cr(VI) Reagents

S.No	Substrate	Product	% Yield	Reagent	Ref.
1			22	t-Butylchromate	[86]
2	 R³ = OCH₂Ph	RR¹ = O; R² = α-OH RR¹ = O; R² = OCH₃	1.2[a] 41[b]	PCC	[90]
3			29	CrO₃	[91]
4	Secododecaherene (**6**)	C₂-Symmetric diepoxide(**7**)		Na₂CrO₄–Ac₂O–Acetic acid	[33]

(continued)

Table 3.7 (continued)

S.No	Substrate	Product	% Yield	Reagent	Ref.
5				$K_2Cr_2O_7$ [c]	[92]
6			45[d]	PFC	[93]
7			5	PFC	[36]

[a] DMF at room temperature for 12 h
[b] Methanol at 35 °C for 13 h
[c] Ethylacetate in presence of NH_3, Ethanol and Ethylacetate
[d] The expected diketone was not at all formed

appears to be a dehydrogenation process, it might have been proceeded through a normal alcohol oxidation followed by tautomerisation as depicted in Scheme 3.4.

Other interesting results of such unusual course of reactions are listed in Table 3.7. The kinetic and mechanistic details of these reactions are yet to be explored and may lead to interesting revelations.

References

1. Narayanan N, Balasubramanian TR (1986) Indian J Chem 25B:228
2. Firouzabadi H, Sardarian A, Gharibi H (1984) Synth Commun 14:89
3. Ruggero Curci, Luigi Lopez, Luigins Troisi (1988) Tetrahedron Lett 29(25):3145
4. Santanielo E, Milani F, Casati R (1983) Synthesis 749
5. Gun LeeJong, Soo HaDong, Jong LimHee (1988) Bull Korean Chem Soc 9B:407
6. Harry DB, Roger SM, Jeanne BM, William PW (1983) Heterocycles 20:2029
7. Frank S Guziec Jr, Frederick A Luzzio (1980) Synthesis 691
8. Bhattacharjee MN, Chaudhui MK, Dasgupta HS, Roy N, Khathing DT (1982) Synthesis 588
9. Frechet Jean MJ, Darling P, Farrall Jean MJ (1981) Org Chem 46:1728
10. Santaniello E, Ferraboschi P, Sozzani P (1980) Synthesis 646
11. Balasubramanian K, Prathiba V (1986) Indian J Chem 25B:326
12. Lopez CM, Gonzalez A, Cossio FP, Palomo C (1985) Synth Commun 15:1197
13. Corey EJ, Suggs JW (1975) Tetrahedron Lett 2647
14. Tang M, Huang X, Chen Z, Hangzhou daxue X (1984) Ziran Kexucban 11(4):454. CA:102:148847b
15. Huang X, Chan C-C (1982) Synthesis 1091
16. Corey EJ, Schmidt G (1979) Tetrahedron lett 399
17. Ozgun B, Degrimenbasi N (1996) Synth Commun 29:3601
18. Ion S, Adalgiza C, Anca N, Felicia C (1983) Rev Chim (Bucharest) 34:588
19. Goffredo R, Roberto B (1983) Synthesis 543
20. Ingle TR, Dhekne VV, Kulkarni VR, Rama Rao AV (1983) Ind. J. Chem 22B:69
21. Czerneki S, Georgoulis C, Stevens CL, Vijayakumaran K (1986) Synth Commun 16(1):11
22. Anjaneyalu ASR, Sastri CVM, Umasundari P, Satyanarayanan P (1987) Ind J Chem 26B(14):305
23. Corey EJ, Samuelson B (1984) J Org Chem 49:4735
24. Francisco R, Alberto G, Claudo P (1986) Carbohydr Res 149:C1–C4
25. Jozef J, Ryszard N (1987) Pr. Nank. Politech. Szezecin., 329:21 CA:109:37592q
26. Sternback DD, Rossana DM (1982) Tetrahedron Lett 23:303
27. Santaniello E, Ferraboschi P (1980) Synth Commun 10:75
28. Guziec Jr FrankS, Luzzio FrederickA (1982) J Org Chem 47:1787
29. Aizpurua JM, Claudio P (1983) Tetrahedron Lett 24:4367
30. Battacharjee Manabendra N, Chaudhuri Mihir K, Subrata P (1987) Tetrahedron 43:5389
31. Bhat NB, Bhaduri AP (1985) Ind J Chem 24B:1286
32. Jong Gun L, Hee Jong L, Dong Soo H (1987) Bull Korean Chem Soc 8(5):435
33. Rajasekhar N, Subramaniam R, Gould ES (1982) Inorg Chem, 21:4110; and (1983), 22, 971
34. Francesco B, Romano DF, Carlo B (1984) J Org Chem 49:1647
35. Radhika R, Eisenbraun EJ (1985) J Org Chem 50:2435
36. Varadarajan R, Dhar RK (1986) Indian J Chem 25B:971
37. Walter K, Dieter M (1984) Z Chem 24(5):182
38. Michio S, Tetsutaro I (1983) Kanagawa Daigaku Kogakubu Kenkyu Hokoku, 21 :36. CA :99:157788e
39. Brunelet T, Gelband G (1985) J Chem Res Synop 264

40. Vangalur S, Srinivasan (1982) Inorg Chem 21:4328
41. Lianquan G, Benli L, Jingling Z (1985) Faming Zhuahli Shenqing Gongkai Shuomingshu CN., 85:102 019. CA:106:32573n
42. Kinji H, Kiyoto G, Yoshiaki T (1985) Japan Kokai Tokkyo Koho JP, CA:104:50694r
43. Bose RN, Gould ES (1986) Inorg Chem 25:94
44. O'Connor B, George J (1987) Tetrahedron Lett 28:3235
45. Brown HC, Chandra GP (1986) Tetrahedron 42:5511
46. Krisna RR, Chawla HPS, Dev Sukh (1984) Indian J Chem 23B:1190
47. Kieth BD, Amonda PJ, Norman T, Derek MG (1986) Fuel Process Technol 14:183
48. Shamsuzzaman S, Ansari JA (1988) Indian J Chem 27B:676
49. Moussa GEM, Salem MR, Shaben ME, Nagdi EI (1987) J Chem Soc Pak 9:391
50. Sarawadekar RG, Menon AR, Banker NS (1983) Thermochim Acta 70:133
51. Cacchi S, La Torra F, Misti D (1979) Synthesis 356
52. Nalliah C, Strickson JA (1986) Tetrahedron 42:4083
53. Nalliah C, Strickson JA (1986) Tetrahedron 42:4089
54. Grigor'ev MG, Buketova IA, Polevschikov PF (1987) Izv.Vyssh.Uchebn. Zaved. Khim. Technol, 30:22 Chem Abstr, 109:6142p
55. Fatidi AJ (1976) Synthesis 65
56. Fried J, Edwards JA (1972) Organic Reactions in Steroid chemistry, Vol. 1, Van Nostrand reinhold Co., New York 244
57. House HO (1972) Modern synthetic reactions. W A Benjamin Inc, Menio Park, CA, pp 265–267
58. Someswara Rao C, Deshmukh AA, Thakor R, Srinivasan PS (1986) Indian J Chem 25B:324
59. Walba David M, DePuy Charles H, Garbowski Joseph J, Bierbaum Veronica M (1984) Organometallics 3:498
60. Parish EJ, Jr Schroepfer (1980) Chem Phys Lipids 27:281
61. Parish EJ, Sarawanee C, Susan L (1984) Lipids 19:550
62. Parish EJ, Scott AD (1983) J Org Chem 48:4766
63. Parish J, Scott AD, Dickerson JR, Dykes W (1984) Chem Phys Lipids 35:315
64. Herscovici J, Marie jose E, Antonakis K (1982) J.Chem. Soc. Perkin Trans 1:1967
65. Parish EJ, Yi Wei Tsao (1987) Synth Commun 17:1227
66. Ifzal SM, Rehena Ahmed, Irfan-ul-haque (1988) Pak J Pharm Sci 1(1):21
67. Rihter B, Masnovi J (1988) J Chem Soc Chem Commun 35
68. John GT, William KJ Jr (1985) Isr J Chem 25:148
69. Alexander K, Jan R, Hans Dieter S (1984) Liebigs Ann Chem, 783. CA : 100 : 210294 h
70. Khan Mushfiquddin, Ahmad S, Osman SM (1986) Indian J Chem 25B:321
71. Singh Chandran (1985) Indian J Chem 24B:300
72. Parish EJ, Chitrakorn S (1985) Synth Commun 15:393
73. Holecek J, Handlir K, Nadvornik M (1983) J Prakt Chem, 325(2):341; Chem. Abstr., 99:175981 s
74. Parish EJ, Yi Wei Tsau, Peter L (1987) Lipids 22:760
75. Francesco Bonadies, Carlo Bonini (1988) Synth Commun 18(13):1573
76. Reddy PS, Yadagiri P, Lumin S, Shin DS, Falck JR (1988) Synth Commun 18(5):545
77. Schlecht Mathew F, Jin KimHo (1989) J Org Chem 54(3):583
78. Maeorg U, Saar M, Kokk H, Paama L (1988) Otkrytiya. Izobret, 30:87. C.A. 110:175504j
79. Wu Hsien Jen, Pan Kai, Hsia Ket Shang, Liu Wei Chi (1988) J.Chi. Chem. Soc. (Taipei),), 35(3):227. C.A. 110:173020z
80. Roberto A, Maurizio DA, Antonella DM, Giovanni P, Arrigo S (1983) Tetrahedron 39:1765
81. Khan Mushfiquddin, Agawal R, Ahmad S, Ahmad F, Osman SM (1988) Indian J Chem 27B:430
82. Czernecki S, Georgoulis C, Sevens CL, Vijayakumaran K (1985) Tetrahedron Lett 26:1699
83. Drabowicz J (1980) Synthesis 125
84. Hsing Jang L, In Sup H (1985) Synth Commun 15:759
85. Corey EJ, Ernie Paul B, Magriotis PA (1985) Tetrahedron Lett 26:5855

86. Roberts DonaldL, Heckman RobertA, Hege BonitaP, Bellin StuartA (1968) J Org Chem 33:3566
87. Varadarajan R, Dhar RK (1986) Indian J Chem 25B:746
88. Paquette LA, Kobayashi T (1987) Tetrahedron Lett 28:3531
89. Anna BM, Rodney RW (1983) Org Prep Proceed Int 15:239
90. James HR, Peter HB, Yang-Zhi L, Jia-Yi T (1985) J Chem Res Synop 46
91. Hussan F (1987) Aust J Chem 40:999
92. Nicholas NR, David QA (1983) Synth Commun 13:817
93. Srinivasan Vangalur S, Gould ES (1981) Inorg Chem 20:3176

Chapter 4
Kinetic Aspects

Abstract In this chapter, the general kinetic characteristics such as the order, activation parameters and isotopic effect on the rate of oxidation with some of the Cr(VI) reagents and their mechanistic implications are discussed.

Keywords Kinetics · Oxidation · Activation parameters · Isotope effect · Mechanism

4.1 General Aspects

Oxidations with Cr(VI) reagents are, generally, smooth and clean and proceed with conveniently followable velocity at ambient temperatures. The syntheses involving these oxidants have been carried out mostly in less polar solvents such as dichloromethane and trichloromethane, though the majority of kinetic measurements have been made only in polar solvents like aqueous acetic acid. However, a peculiar solvent composition consisting of equimolar mixture of acetonitrile and nitrobenzene has been employed [1, 2] in reactions with PFC and PCC. The frequently encountered stoichiometric ratio in the oxidation of alcohols with Cr(VI) reagents is 3:2. Usually, the order with respect to the substrate and oxidant is each unity; though, deviations have been observed in some reactions (Table 4.1). The oxidizing potency of dichromates with different inorganic counter ions has been compared in a report [3] and found to be $K_2Cr_2O_7 < Na_2Cr_2O_7 < Li_2Cr_2O_7 < (NH_4)_2Cr_2O_7$. In the chromic acid oxidation of mercaptans, surprisingly a fractional order dependence on oxidant has been reported [4]. Similarly, in the oxidation of glycolic acid by chromium peroxidichromate, a 'pseudo zero order' dependence on the concentration of substrate has been reported [5]. The oxidation of lactic acid exhibits [6] a variable order dependence on the substrate. Chromic acid oxidations are usually catalysed by mineral acids

S. Sundaram and P. S. Raghavan, *Chromium-VI Reagents: Synthetic Applications*, SpringerBriefs in Molecular Science, DOI: 10.1007/978-3-642-20817-1_4, © P. S. Raghavan 2011

Table 4.1 Kinetic characteristics of oxidations by Cr(VI) reagents

S.No	Substrate(S)	Oxidant(Ox)	Experimental conditions	Order on [S]:[Ox]	Ref
1	Glycolic acid	Chromium peroxydichromate		0:1	[5]
2	Formic acid	Chromium peroxydicromate		2:1	[20]
3	Ethanol; Benzyl alcohol; Cyclohexanol	PFC	Acetonitrile–Nitrobenzene (1:1); 303 K	0:1	[1]
4	Benzhydrol	PCC	DMSO	1:1	[21]
5	L – Methionine	Chromic acid	$HClO_4$:25 °C	1:1	[22]
6	Citric acid	Chromic acid	Acid medium		[23]
7	Oxalic acid	Chromic acid	Acid medium		[24]
8	Indigocarmine	Chromic acid	$HClO_4$	1:1	[25]
9	D-Glucopyranose-6-phosphate;	Chromic acid	$HClO_4$	1:1	[26]
10	CH_3OH; CH_2DOH	Chromic acid	$HClO_4$	1:1	[7]
11	Acrylic acid	Chromic acid	$HClO_4$–HNO_3		[27]
12	Butanol	Chromic acid	$HClO_4$	1:1	[28]
13	R-SH (R = methyl, propyl)	Chromic acid	Aqeous acetic acid	1:x x < 1	[4]
14	Lactic acid	Chromic acid	Aqueous medium	Variable*	[6]
15	HI	Chromic acid	Aqueous medium		[29]
16	D-Glucose	PFC		1:1	[30]
17	Phenylmethylsulphide	PCC	Acetonitrile–Nitro benzene(1:1);	x:1 x < 1	[2]
18	Phenylmethylsulphide	PCC	60% aq. Acetic acid	1:1	[31]
19	DMSO	Chromic acid	Aqueous medium	1:1	[8]

* Order varies from 0 to 2 depending upon the concentration of the substrate

and the order with respect to $[H^+]$ is very often either one or two. However, in some exceptional cases a third or fourth order dependence have also been reported [7, 8]. Many other interesting observations including an inverse order dependence on the Cr(VI) oxidant concentration in the oxidation of acetophenoxime [9], a shift of order dependence on the substrate concentration for homologous substrates from second order on oxalic acid and first order on malonic acid [10], co-oxidation of aliphatic carboxylic acids [11–15], etc., have also been reported. Besides, routine kinetic reports on many systems like glycols [16], thioglycols [17] and glycolic acid [18] by dichromate and benzyl alcohols [19] by PFC have appeared in literature.

There had been very interesting arguments and counter arguments in the past, on the nature of the transition state of the chromic acid oxidation of several substrates, alcohols in particular, based on the results obtained from the application of Hammett free energy relationship. However, only a limited number of reports are available on the structure—activity relationship in reactions (Table 4.2) involving novel Cr(VI) reagents. For all the substrates, only a negative value for

Table 4.2 Hammett correlations in oxidations by Cr(VI) reagents

S.No	Substrate	Oxidant	ρ	Ref
1	Toluenes	PFC	-2.00	[32]
2	Phenylmethylsulphides	PCC	-2.12^{a}	[2]
3	Phenylmethylsulphides	PCC	-1.23^{b}	[2]
4	Arylalkylsulphides	$K_2Cr_2O_7$	-2.10	[33]
5	Diphenylsulphides	$K_2Cr_2O_7$	-1.80	[33]

[a] in protic solvent; [b] in aprotic solvent

the Hammett's reaction constant rho (ρ) has been obtained indicating the electron deficient nature of the transition state. The values for the ρ, in general, are all about -1.0 to -2.2 in any protic or aprotic solvents (Table 4.2).

The activation parameters for the oxidation of a variety of substrates by different Cr(VI) reagents have been evaluated by different group of research workers [1, 7, 32, 34, 35]. The sign and the relative magnitude of these values would help in deciding the nature and the topology of the transition state involved in the reaction. However, it is seen that a vast majority of the values for the activation energy, E_a fall in the range 50–60 kJ/mol and those of the free energy (ΔG) converging to 92–94 kJ/mol. This apparent insensitivity of these activation parameters to the change in the nature of substrate or oxidant may be due, at least partially, to the aqueous acetic acid solvent system used in many of these cases. In this solvent medium, since the reagent may potentially ionize before the attack, the species generated from all the reagents may be identical.

4.2 Isotope Effects

The discovery of deuterium in 1932 and the in-depth knowledge about the zero point stretching vibrations of C–H and C–D bonds added one more dimension–isotope effects—to the kinetic study of reactions. Books and monographs are available [36–38] providing a detailed picture of the fundamental basis of isotope effects on organic reactions. The decrease in zero point stretching level that accompanies deuteration can be readily determined for virtually any organic compound since, the vibrational frequency v_{C-D} occurs in the range 2100–2300 cm^{-1} in the infra red, a region of the spectrum that is otherwise, generally, transparent. The assumption that the stretching mode becomes a translational mode in the transition state and that no other factor enters, leads to expected maximum values of k_H/k_D in the range 6–7 at room temperature. Lower values of k_H/k_D have usually been rationalized in terms of the transition state acquiring an unsymmetrical stretching mode which reduces the zero point energy difference between reactants and transition state [39–41] or in terms of the effects on bending modes. On the other hand, higher k_H/k_D values are usually explained in terms of quantum mechanical tunneling [42] although bending effects have also been considered.

Table 4.3 Primary Kinetic Isotopic Effect (PKIE) Data

S.No	Substrate	Oxidant	k_H/k_D	Ref
1	Naphthalene/Naphthalene-d_8	PFC	5.50	[34]
2	p-$CH_3C_6H_4CH_3$/p-$CH_3C_6H_4CD_3$	PFC	5.40	[32]
3	m-$CH_3C_6H_4CH_3$/p-$CH_3C_6H_4CD_3$	PFC	5.40	[32]
4	o-$CH_3C_6H_4CH_3$/p-$CH_3C_6H_4CD_3$	PFC	5.10	[31]
5	Ethanediol/Ethanediol-1,1,2,2-d_4	PCC	6.75	[44]
6	CH_3OH/CH_3OD	Chromic acid	1.00	[7]

The maximum value of k_H/k_D is also attributed to a centrosymmetric transition state [39, 43] and any deviation from the maximum value rationalized on the basis of a reactant—like or a product-like transition state.

A scrutiny of the reported results on kinetic isotope effects (Table 4.3) is interesting. The report on k_H/k_D values in naphthalene and xylene system is puzzling, as the authors themselves [32, 34] claim that the reaction is almost independent of the substrate concentration. Even granting a small fractional order on substrate the acceptability of the directly determined k_H/k_D value is questionable. It is also reported by other group of workers [7] that the oxidation of methanol and its deutrated analogue, CH_3OD, by chromic acid renders a value of $k_H/k_D = 1$. It is not unexpected, since it is well known that in the oxidation of alcohols that only a C–H bond breaks in the rate determining step of the reaction and the intension of using the O–D compound is eluding.

A more detailed understanding of the nature and structure of the transition state and hence its complete topological perspective can be had by studying the temperature dependence of kinetic isotope effect (TDKIE) of a reaction [45]. The TDKIE criterion as a tool to unravel the topology of transition state is based on the Arrhenius relationship:

$$k_H/k_D = (A_H/A_D) \exp\left(-[\Delta Ea]_D^H/RT\right)$$

where A_H and A_D are the respective frequency factors and $[\Delta Ea]_D^H$ is the difference between the activation energies of the protiated and deuterated compounds. Depending upon the magnitude of the factors A_H/A_D and $[\Delta Ea]_D^H$ referred to as the TDKIE parameters, the transition states involving hydrogen transfer can be classified into four types and the salient features are summarized in Table 4.4.

4.3 Mechanistic Implications

Mechanistic investigations on the oxidation of a variety of substrates including carbohydrates [46, 47], carboxylic acids [20, 22, 48, 49], arylalkyl sulphides [33, 50], hydrocarbons [28, 34], ethers [51, 52], phenols [53], sulphoxides [54, 55], oxamic acid [56], mercaptans [4], acetals [57, 58], α-hydroxyacids [6, 57–61] and esters [62] by different Cr(VI) oxidants are abundant in literature.

Table 4.4 Classification of transition states based on their TDKIE parameters[*]

S.No	Transition state category	k_H/k_D	$[\Delta E_a]^H_D$ [k cal]	A_H/A_D
1	Linear H—transfer; Symmetrical	Maximum (6–8 at 25 °C)	$= [\Delta E_a]^H_D$ 1.15[†]	0.7–1.414
2	Linear H—transfer; Unsymmetrical	<Maximum (2–5 at 25 °C)	$\leq [\Delta E_a]^H_D$ 0.3–1[†]	0.7–1.414
3	Linear H—transfer; Tunnelling	>Maximum (> 9 at 25 °C)	$\gg [\Delta E_a]^H_D$ 1.5–6[†]	<0.2
4	Non-Linear H—transfer; Bent	>2	0	2–6

[*] Data taken from Ref. [45]
[†] For C—H(D) bonds

The generally accepted mechanism for the alcohol oxidation by chromic acid envisaging the chromate ester formaton and its slow decomposition to products could be extended to the oxidation by other chromium(VI) species as well (Scheme 4.1).

$$R_2CHOH + Q^+CrO_3X^- + H^+ \underset{}{\overset{\text{fast}}{\rightleftharpoons}} R_2CHOCrO_2X + Q^+ + H_2O$$

$$R_2CHOCrO_2X \xrightarrow{\text{slow}} R_2C{=}O + HCrO_2X$$

Scheme 4.1

The rate determining step involving a C–H cleavage has been demonstrated with experiments on kinetic isotope effects. The mode of hydrogen transfer (H^+ or H^-) from carbon to oxygen has been disputed and kinetic parameters, particularly the Hammett rho value or the catalysis by bases have been frequently quoted in support of electron rich and electron deficient nature of the departing hydrogen respectively [42, 63–65] (Scheme 4.2a, b).

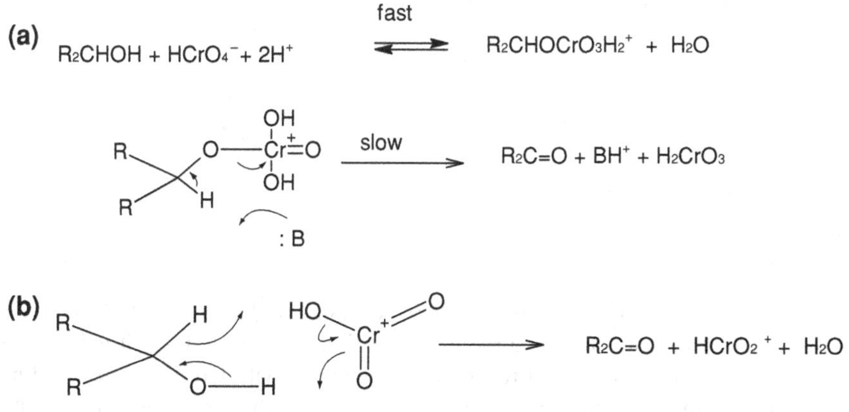

Scheme 4.2

Besides a unimolecular decomposition of the ester by internal proton transfer has also been suggested [66] (Scheme 4.3).

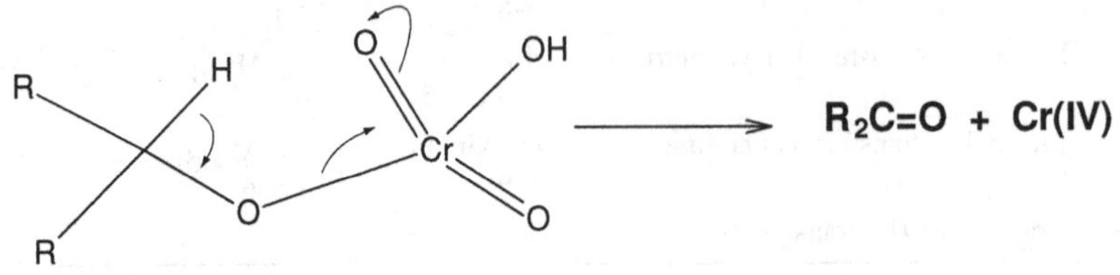

Scheme 4.3

A hydrogen atom transfer has been proposed [32, 34] for the oxidation of aromatic hydrocarbons by PFC (Scheme 4.4).

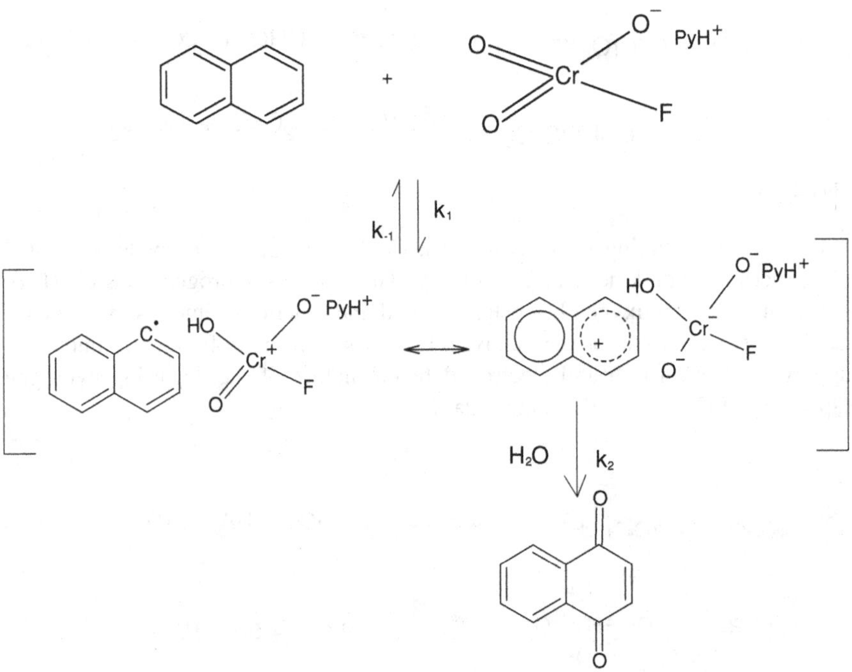

Scheme 4.4

It is interesting to note that though these reactions (Scheme 4.4), fail to induce polymerization of acrylonitrile, a radical formation is suggested and it is believed that the initially formed radical reacts instantaneously with Cr(V) species leading to the carbonium ion. The carbonium ion character in the transition state is substantiated by a large negative reaction constant.

Surprisingly, Cr(VI) fails to oxidize Tl(I) in presence of either H_2SO_4 or $HClO_4$ but the reaction is facile when HCl is used [67, 68]. This differential behavior in the presence of acids has been ascribed to the chlorochromate formed in situ being the active oxidant rather than the mere Cr(VI) species in the reaction.

On the contrary, a similar in situ generation of chlorochromate species has been invoked to account for the rate reduction by the added NaCl in the oxidation of acetals by chromic acid [58]. The mechanistic details of oxidation of several substrates with Cr(VI) oxidants other than the simple dichromate have been reported in literature [1, 32, 44, 50]. However, the real impact of these modifications on the rate has evaded detection because the acidic nature of the chosen solvent medium often swamps other discriminating factors. Ruthenium (IV) catalysed oxidations of alkylbenzenes [69] by Cr(VI) and oxidationsby CrO_3 supported on Celite or florosil, of nitrosoamines and nitrosoureas have been documented [70]. Cr(VI) complexes have also been used on support and in solution for oxidation of inorganic hydrides [71]. Oxidations at liquid–liquid [72] and liquid–air [73] interfaces of aldehydes and alcohols respectively by chromic acid constitute a new branch of development. The rate of oxidation at the liquid–liquid interface is higher than that in bulk and the liquid–air interfacial oxidation studies uphold the C–H bond breaking in the rate determining step.

References

1. Battacharjee Manbendra N, Chaudhuri Mihir K, Dasgupta Himadri S (1984) Bull Chem Soc Jpn 57:258
2. Rajasekharan K, Bhaskaran T, Gnanasekaran C (1984) J Chem Soc Perkins Trans 2:1183
3. Grigor'ev MG, Buketova IA, Polevschikov PF (1987) Izv Vyssh Uchebn Zaved Khim Technol 30:22 Chem Abstr 109:6142
4. Nagi Reddy K, Uma Devi PS, Saiprakash PK (1983) Acta Cienc Indica Chem 9:112. Chem Abstr 100:208915t
5. Valechha ND, Dakwala DS (1984) J Indian Chem Soc 61:581
6. Samal PG, Patnaik BB, Rao SCD, Mahapatra SN (1983) Tetrahedron 39:143
7. Sen Gupta KK, Tapashi S, Sumarendra Nath B (1985) Tetrahedron 41:205
8. Venkateswra Rao S, Jagannadham V (1988) Indian J Chem 27A:252
9. Rao LV, Prasada S, Brahmaji S (1988) J Indian Chem Soc 65(6):404
10. Anis SS, Mansour MA, Stefan Shaker L (1988) Microchem J 37(3):345
11. Farize Hasan, Rocek J (1976) J Amer Chem Soc 98:6574
12. Ramesh S, Rocek J, Schoellor DA (1978) J Phys Chem 82:2752
13. Ip D, Rocek J (1979) J Amer Chem Soc, 101:6311 and references cited therein
14. Nagarajan K, Sundaram S, Venkatasubramanian N (1979) Ind J Chem 18:260
15. Nagarajan K, Sundaram S, Venkatasubramanian N (1979) Ind J Chem 18A:335
16. Alvarez Macho MP (1988) An Quim Ser A 84(1):14
17. Baldea Ioan (1987) Stud Univ Babes-Bolyai Chem 32(2):42 CA 110:56881v
18. Szeto May Nin-Ping (1988) Diss Abstr Int B 48(10):2979 CA 110:7407 m
19. Banerji KK (1988) J Org Chem 53(10):2154
20. Sharma VK, Rai RC (1983) J Indian Chem Soc 60:745
21. Banerji KK (1983) Indian J Chem 22B:413
22. Adegboyega Olatunji M, Adefikayo Ayoko G (1988) Polyhedron 7:11

23. Alvarez Macho MP, Ontequi martin MI (1987) An Quim Ser A 83:248 CA:108:132217j
24. Crespo Pinilla MC, Mata Parez F, Alvarez Macho MP, Crespo Pinilla (1987) An Quim Ser A 83:130 CA:108:130832g
25. Kamannarayana P, Raghavachari K (1986) Acta Cienc Indica Chem 12:79 CA:108:5430y
26. Sen Gupta KK, Sen Gupta S, Subrata Kumar M, Sumarendra nath B (1985) Carbohydr Res 139:185
27. Alvarez Macho MP, Sermentero Calzada J (1984) An Quin Ser A 80:687 CA:103:27988k
28. Alvarez Macho MP (1984) An Quim Ser A 80:672 CA 103:27986h
29. Uma P, Kanta Rao P, Sastri MN (1985) Indian J Chem 24A:539
30. Varadarajan R, Dhar RK (1986) Indian J Chem 35A:474
31. Herscovici J, Marie Jose E, Antonakis K (1982) J Chem Soc Perkin Trans 1:1967
32. Bhattacharji B, Bhattacharji MN, Bhattacharji M, Bhattachaji AK (1986) Bull Chem Soc Jpn 59:3276
33. Srinivasan C, Chellamani A, Rajagopal S (1985) J Org Chem 50:1201
34. Bhattacharji B, Bhattacharji MN, Bhattacharji M, Bhattachaji AK (1985) Int J Chem Kinet 17:629
35. Banerji KK (1978) J Chem Soc Perkin Trans 2:639
36. Bell RP (1973) The proton in chemistry. Cornell University Press, Ithaca, NY
37. Melander L (1960) Isotope effects on reaction rates. Ronald Press, NY
38. Wiberg KB (1955) Chem Rev 55:713
39. Westheimer FH (1961) Chem Rev 61:265
40. Bell R P (1965) Discuss Faraday Soc 16
41. Albery WJ (1967) Trans Faraday Soc 63:200
42. Stewart R, Lee DG (1964) Can J Chem 42:439
43. Venkatasubramanian N (1965) Indian J Chem 3:225
44. Banerji KK (1983) Indian J Chem 22B:650
45. Kwart H (1982) Acc Chem Res 15:401
46. Sharma K, Sharma VK, Rai RC (1983) J Indian Chem Soc 60:747
47. Raj Rabindra N, Glushka John N, Chew Warren, Perlin Arthur S (1989) Carbohydr Res 189:135
48. Manorama S, Panigrahi GP, Mahapatra SN (1985) J Org Chem 50:3651
49. Rao Venkateshwar G, Saiprakash PK (1988) Oxid Commun 11:33
50. Rajasekharan K, Bhaskaran T, Gnanasekaran C (1987) Indian J Chem 26A:956
51. Musale Reddy P, Jagannadham V, Sethuraman B, Sethuraman B, Navaneeth Rao T (1982) Pol J Chem 56:865
52. Chidambaram N, Sathyanarayan K, Chandrasekaran S (1989) Tetrahedron Lett 30(18):2429
53. Satyanarayana Reddy T, Jagannadham V (1986) Proc Natl Acad Sci India 56A:129
54. Bhatta D, Behera SM, Mohanty SR (1987) Indian J Chem 26A:778
55. Balaiah V, Satyanarayana PVV (1985) Curr Sci 54:384
56. Kharga RD, Mishra HB (1983) J Nepal Chem Soc 3:45
57. Nambi K, Basheer AKA, Arulraj SJ (1988) J Indian Chem Soc LXV(2):85
58. Basheer AKA, Nambi K, Arulraj S (1987) Indian J Chem 26A:672
59. Saran NK, Acharya RC, Rao SR (1985) J Indian Chem Soc 62:747
60. Height GP, Gregory GM, Terry KM, Patrick M (1985) J Inorg Chem 24:2740
61. Singh VP, Fandey IM, Sharma SB (1985) J Indian Chem Soc 62:64
62. Gupta KK, Taneja SC, Dhar KL (1989) Indian J Chem B 28B(2):136
63. Wetheimer FH, Leo A (1952) J Am Chem Soc 74:4383
64. Rocek J, Krupicka J (1958) Coll Czech Chem Commun 23:2068
65. Venkatasubramanian N (1960) Proc Indian Acad Sci 50:156
66. Kwart H, Francis PS (1959) J Am Chem Soc 81:2116
67. Gokavi GS, Raju R (1988) Indian Chem 27A:494
68. Sagi R, Prakasa Raju GS, Ramana KV (1975) Talanta 22:93
69. Tret'yakov VP, Zimtseva GP, Min'ko LA, Chudaev VV, Rudakov ES, Tyupalo NV (1987) Ukr Khim Zh 53(11):176 C.A.:109: 189624q

70. Saavedra JE, Fransworth W, Guokui Pei (1988) Synth Commun 18(3):313
71. Dorfman Ya A, Polimbetova GS, Kelman IV et al (1989) Koord Khim 15(1):77 C.A. 110:180124q
72. Mimicos N, Mylona A, Paleos CM (1987) Mol Cryst Liq Cryst 543
73. Ahamed Jamil Astin, Brain K (1988) Langmuir 4(3):780

References

Chapter 5
Solid Supported Cr(VI) Reagents

Abstract The discussion in the previous chapters were mainly restricted to the applications of Cr(VI) reagents in homogeneous media. However, some of the Cr(VI) reagents have also been found to exhibit much desired reaction characteristics such as enhanced yield of the product, reduced reaction time, easy handling, simple method of separating the reaction product with higher purity, etc., when these reagents are anchored on to a suitable solid support. This chapter is intended to give a brief idea of the general merits and demerits of solid supported reagents and in particular, the preparation of certain solid-supported Cr(VI) reagents and their applications as oxidants in heterogeneous conditions are discussed.

Keywords Solid support · Cr(VI) reagent · Polymer support · Immobilised oxidant · Oxidation

The reagents considered earlier are all very useful Cr(VI) oxidizing agents mostly employed in homogeneous solution phase. Consequently, the isolation of the products becomes laborious and might involve multi step processes. These difficulties could be overcome to an extent by employing oxidizing agents anchored onto a solid support. Several such solid supported chromium (VI) oxidants have been developed and their oxidizing potential evaluated. Such solid supported reagents have been reported to modify the activity of the reagent, improve the selectivity, and of course, make product isolation much easier. In many reactions, the performance of solid supported Cr(VI) reagents have been found to be superior to the non-supported reagent. The main advantages of these solid supported reagents include the following:

1. They can be easily removed from reactions by filtration.
2. Excess reagents can be used to drive reactions to completion without introducing difficulties in purification.
3. Recycling of recovered reagents is economical and hence environmentally-friendly and efficient.

S. Sundaram and P. S. Raghavan, *Chromium-VI Reagents: Synthetic Applications*, 57
SpringerBriefs in Molecular Science, DOI: 10.1007/978-3-642-20817-1_5,
© P. S. Raghavan 2011

4. Usually, reagents on solid-support react differently, mostly more selectively, than their unbound counterparts.

However, there are some issues as listed below, which may restrain the choice of solid-supported reagents for certain specific reactions.

1. Some reagents may not interact well with the solid support.
2. Reactions on solid support may be slow due to diffusional constraints.
3. Polymeric support materials can be very expensive to prepare.
4. Stability of the support material can be poor under harsher reaction conditions.

The preparation and the applications of some of the solid supported Cr(VI) reagents are discussed in this section.

5.1 Cr(VI) Reagents Supported on Alumina

CrO_3 is a mild but efficient and inexpensive reagent. However, when it is used as an oxidant in acid medium for alcohols, often, it forms carboxylic acid and some times leaves tarry materials. When it is loaded on alumina and employed in reactions, generally reactions are very clean and it converts alcohols to aldehydes without any further oxidation to carboxylic acids and also no tar is formed. The deprotection of trimethylsilyl and tetramethylpyranyl ethers by CrO_3-supported on alumina has been reported [1] to be quick and efficient with the reaction times ranging from 30 min to 3 h and the product yield 72–90%, depending on the nature of the ether. The reagent is prepared by thoroughly mixing finely powdered chromium oxide with wet alumina powder and the ethers taken in dichloromethane solvent being run over the bed of this reagent. A little wet condition of the reagent is required for the reaction as the reaction does not proceed with dry reagent. Quinolinium fluorochromate supported on alumina has been found to be highly stable and selective. It was also reported to be effective reagent for the oxidation of alcohols and oxidative deoximation of aldoximes and ketoximes [2, 3].

The acidic nature of PCC could be reduced by adsorbing it onto alumina. When primary alcohols are oxidized with alumina supported PCC under solvent-free conditions, the substrate is directly converted to the corresponding alkyl alkanoate [4]. On the other hand, the reaction under similar conditions with benzylic primary alcohols and allylic alcohols resulted in the formation of the conventional aldehydes as products and secondary aliphatic and aromatic alcohols yielded ketones.

With molecules containing both the aliphatic and aromatic hydroxyl groups, a cyclic product is also obtained besides the alkyl alkanoate. In general, the alumina supported PCC is preferred as it does not induce any isomerisation of the ketonic products formed or polymerization of the double bonds present in the substrates; also, no side reactions and/or over oxidation to carboxylic acids occur.

3-Carboxypyridinium chlorochromate (3CPCC) supported on alumina has been shown to be a good oxidant for primary and secondary alcohols. The efficiency of this oxidant manifolds when the reactions are conducted under microwave irradiaton [5].

5.2 Cr(VI) Reagents Supported on Silica Gel

Jone's reagent supported on silica gel can be very easily prepared by slow addition of about 1.5 mL (12 mmol) of Jone's reagent to a vigorously stirred dry silica gel (5 g) sample. The stirring should be continued for about five minutes after the addition is completed, when a free flowing orange powder of the required reagent is obtained. The advantages of this reagent is that the activation of the reagent prior to use is not required and that the amount of chromium(VI) oxidant present on the silica gel is precisely known. This reagent has been shown [6, 7] to almost quantitatively (98–99%) oxidize benzyl alcohol and substituted benzyl alcohols to the corresponding aldehydes.

$$R-C_6H_4-CH_2-OH \rightarrow R-C_6H_4-CHO$$

where R=2–CH$_3$, 2–Cl, 4–CH$_3$, 4–Cl, 4–NO$_2$; It is also used efficiently in the transformation of secondary alcohols to ketones, oxidative regeneration of carbonyl compounds from oximes, oxidation of aldehydes to carboxylic acids, oxidation of hydroquinones to quinines and sulfides to sulfones.

Cr(VI) ions adsorbed on certain binary oxide systems such as silica-ziconia have also been developed. Lucia et al. [8] prepared one such system by using ZrO$_2$ coated on the surface of silica gel to absorb Cr(VI) ions from acidic solutions. This oxidant material, in conjuction with t-butylhydroperoxide proved to be very useful for allylic oxidations. The reactions are also very clean and are highly regioselective.

Quinolinium fluorochromate [9] supported on silica gel is so selective that it facilititates the oxidation of axial hydroxyl rather than the one at the equatorial position of a cyclohexane ring. For example, when the *cis* and the *trans* isomers of t-butyl cyclohexanols were independently subjected to oxidation by QFC on silica gel, the axial hydroxyl group of the cis-isomer was oxidized in 2 h with 92% yield while the equatorial hydroxyl group of the trans-isomer yielded 63% product in 4.5 h [10]. (Scheme 5.1).

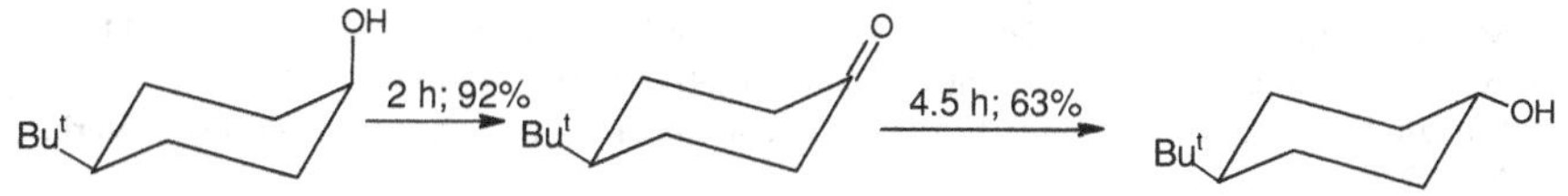

Scheme 5.1

PCC adsorbed onto silica gel has been reported to be a better oxidant than the simple reagent in solution for the oxidation of organic compounds [11]. This reagent is so efficient that it oxidizes aromatic and cinnamylic aldehydes to the corresponding acids under solvent-free conditions while these reaction do not occur at all with just PCC in solution! [12]. A remarkable increase in the product yield could be obtained in the oxidation of primary and secondary alcohols when PCC adsorbed onto silica gel is employed as an oxidant under ultrasound conditions [13].

5.3 Polymer Supported Cr(VI) Reagents

Polymer supported reagents are beneficial in terms of their simplicity in operation, filterability and regenerability. Besides, the polymer matrix also provides selectivity in a reaction due to the steric constraints imposed by the macromolecular matrix or certain specific microenvironmental effects. Many such polymer supported Cr(VI) reagents including divinylbenzene, ethyleneglycoldimethacrylate and NN'-methylene bis-acrylamide cross-linked poly(methylmethacrylate)-supported pyrazolium chromate, chlorochromate and pyrazole-CrO_3 complex resins have been reported [14]. These resins have been shown to selectively oxidize primary alcohols to aldehydes and secondary alcohols to ketones.

A very similar divinylbenzene and ethyleneglycol dimethyl acrylate (EGDMA) cross-linked polystyrenes have also been functionalized to generate pyrazolinium chromate, chlorochromate and pyrazole-CrO_3 complex [15]. The EGDMA cross-linked polystyrene supported reagents have been found to show higher reactivity in terms of functional group capacity and product yield. Cr(VI) reagents supported on certain graft copolymers of polypropylene and polyfluoro ethylene with 4-vinyl-pyridine and also of polyacrylonitrile with 2-methyl-5-vinyl pyridine have also been reported [16]. More information on the synthesis and aplications of different varieties of slid supported reagents in organic reactions could be obtained from the recent review [17] by Solinas Antonio and Taddei Aurizio.

References

1. Aizpurua JM, Claudio P (1983) Tetrahedron Lett 24:4367
2. Rajkumar GA, Arabindoo B, Murugesan V (1998) Indian J Chem 37B:596
3. Rajkumar GA, Arabindoo B, Murugesan V (1999) Synth Commun 29:2105
4. Bhar S, Chaudhuri SK (2003) Tetrahedron 59:3493
5. Mohammadpoor-Baltork I, Memarian HR, Bahrami K (2004) Monatsh Chem 135:411
6. Web site: www.Rhodium.ws. Synth Commun (2001) 31(9):1389–1397
7. Cainelli G, Cardillo G (1984) Chromium oxidations in organic chemistry. Springer–Verlag, New York
8. Baptistella LHB, Sousa IMO, Gushikem Y, Aleixo AM (1999) Tetrahedron Lett 40(14):2695

9. Chaudhuri MK, Chettri SK, Lyndem S, Paul PC, Srinivas P (1994) Bull Chem Soc Jpn 67:1894
10. Rajkumar GA, Sivamurugan V, Arabindoo BA, Murugesan V (2004) Indian J Chem 43B:936
11. Luzzio FA, Fitch RW, Moore WJ, Mudd KJ (1999) J Chem Educ 76:974
12. Salehi P, Firouzabadi H, Farrokhi A, Gholizadeh M (2001) Synthesis 15:2273
13. Adams LL, Luzzio FA (1989) J Org Chem 53:2602
14. Abraham S, Rajan PK, Sreekumar K (1996) Proc Indian Acad Sci Chem Sci 108(5):437–443
15. Abraham S, Rajan PK, Sreekumar K (1998) Polym Int 45:271
16. Verba GG, Asanova M, Ya Plugar N, Abduvakhabov AA, Sh Khakimdzhanov B, Musaev UN (1990) Khimiya Prirodnykh Soedinenii 2:252–255
17. Solinas A, Taddei M (2007) Synthesis 16:2409–2453

Chapter 6
Chromium in Biology

Abstract Chromium, though in its native state as a metal is inert biologically it has certain biological role in its other forms. Cr(III), specifically in complexed form with certain ligands has been reported to be an essential trace nutrient. For example, Cr(III) in the form of a complex called the glucose tolerance factor (GTF) is believed to enhance the peripheral actions of insulin. On the other hand, Cr(VI) species have been shown to be toxic and mutagenic. However, there are different and sometimes contradictory views are projected in literature by various authors about the role of chromium species in biosystems. The facts are presented in this chapter for the self assessment of the researchers in this aspect of chromium.

Keywords Chromium · Trace element · Carcinogen · Glucose metabolism · Diabetes · Insulin · Oligopeptide

Chromium is an essential trace nutrient and its biological activity depends on its valence state. Though chromium exhibits oxidation states from 0 to +6, only +3 and +6 states are stable in simple solids. Metallic chromium (valence zero, Cr-0) is inert. Of the other two stable forms, viz: Cr(III) and Cr(VI), chromium-(III) has been claimed to have some therapeutic value. It is said to be involved in the lipid and sugar metabolism. Bio-available Cr-III salts are found in many foodstuffs, especially liver, American cheese, brewer's yeast, wheat germ, meat, fish, fruits and whole grains. Some vegetables such as carrots, potatoes, and spinach are also good sources, as are alfalfa, brown sugar, molasses, and animal fats [1]. The bioavailability of Cr(III) is higher, especially in complexed form and the chromium picolinate complex is used as a nutritional supplement.

Chromium (VI) is a known toxic and mutagenic substance, while chromium (IV) is known for its carcinogenic properties. Chromate salts have also been found to induce allergic reactions in some individuals. Due to these health and environmental issues, restrictions have been imposed on the use of certain chromium compounds in many countries. The carcinogenic action of Cr(VI) is due to its intracellular reduction to Cr(III) [2, 3], and no carcinogenic activity has been ascribed to ingested Cr(III) [4]. However, Cr(III) has not been adequately evaluated in this regard, perhaps because of its limited ability to cross cell membranes.

The ability of Intranuclear penetration and crossing the cell membrane of Cr(VI) being generally higher it is considered to be definitely mutagenic. However, the usefulness of Cr-III as a long-term nutritional supplement is still debatable due to its possible genotoxic effects [5–7] Norseth [8] expressed the view that all forms of chromium, including Cr(III), should be considered human carcinogens, although this view has been challenged by some authors [9, 10].

Chromium (III) salts have been claimed to have some beneficial effects for patients with diabetic or cardiovascular problems. In combination with nicotinic acid and certain amino acids, Cr-III forms a complex called glucose tolerance factor (GTF) [11, 12]. Although the structure of glucose tolerance factor has not been elucidated, it is known to enhance the peripheral actions of insulin, leading to the concept that people with diabetes might benefit from Cr-III supplementation [13, 14]. Those with heart disease or lipoprotein abnormalities might also require Cr-III supplements, in light of reports of beneficial effects on lipid metabolism [4, 14], provided long-term safety concerns can be resolved.

Ingested chromium is poorly absorbed, (e.g., about 1% of Cr-III and 10% of Cr-VI) [15]. When complexed to other ligands such as nicotinic acid and amino acids in glucose tolerance factor, Cr-III is 25-fold more readily absorbed; Lim et al.[16] used the tracer technique using intravenous ^{51}Cr to study the *in vivo* kinetics of injested Cr(III) species and from the results concluded that it circulates largely, bound to transferrin in the blood. Cr(III) is not stored in the liver, spleen, fat, and bone, and more than two thirds of the absorbed ingested dose is rapidly excreted in the urine and bile. Besides, plasma clearance occurs within 8–12 h, and the tissue elimination has a half-life of several days. The normal concentration of Cr(III) in whole blood is 380–580 nmol/L. Most is contained [17] in erythrocytes, plasma levels being only 1.9–5.8 nmol/L. Serum or urine levels do not reflect body stores of chromium [12]. However, urinary excretion of 10 µg/d is usual in the absence of oral supplementation or industrial exposure [15]. The mechanisms underlying the effects of Cr-III on glucose and lipid metabolism are still being elucidated.

For those exposed to chromium industrially (e.g., during welding or plating), Cr-VI is the valence form involved; it generally enters the body from such exposures by inhalation or transdermally and in this form readily crosses cell membranes. Once within the cellular milieu, Cr-VI is rapidly reduced to Cr-III, which has carcinogenic activity only within the cell nuclei. The strongest evidence that Cr-III supplementation might be beneficial in humans comes from studies by Anderson et al.[13] in patients with diabetes and in healthy subjects whose glucose tolerance was impaired experimentally by chromium depletion.

Walter Mertz, identified a niacin-bound chromium cofactor to insulin, which potentiates the effects of insulin and he called it as glucose tolerance factor, or GTF chromium. Studies have shown that chromium and niacin work synergistically, and together, can reduce cholesterol at doses that avoid niacin's side effects.

With a central role in glucose metabolism, chromium nutrition affects energy levels as well as fat and protein metabolism. A chromium deficiency may lead to hypoglycemia, maturity-onset (Type II) diabetes, and elevated cholesterol—a major cardiovascular risk factor.

Since chromium levels are known to decrease with age, while heart and blood sugar problems tend to rise, mounting evidence suggests that a chromium supplement could be beneficial for many people, particularly in light of the deficiency that occurs in the diet. However, The American Diabetes Association [18] was of the view that "chromium supplementation has no known benefit" in patients with diabetes who are not chromium deficient. Thus, different authors express different, and some times contradictory views on the beneficial effects of chromium in the human physiology. It is of significance to mention here that Mertz in his 1993 review cited 4 positive and 4 negative studies on the effects of Cr-III intake on lipoprotein concentrations At present, the most biologically active form of chromium is said to be an oligopeptide, low-molecular-weight chromium-binding substance (LMWCr) [19–24].

LMWCr is found mainly in the liver of mammals, while appreciable quantities are also found in the kidney. Other tissues examined to date have contained at least traces of the material [25]. LMWCr is also found in urine, where it may represent a major portion of urinary chromium [26]. The LD_{50} value for LMWCr is very small when compared to values for other forms of chromium such as chromate or chromic salts, being 134.9 mg/kg of body weight when given to mice by intra-peritoneal injection [26] Size exclusion chromatography gives a molecular mass of 1.48 kDa for bovine liver LMWCr. The oligopeptide has a chromium-to-protein ratio of approximately 4 to 1; amino acid analyses indicate that aspartate, glutamate, glycine, and cysteine are present in a 2.15:4.47:2.47:2.35 molar ratio, respectively, assuming a molecular mass of 1200 for the organic portion. No significant amounts of other amino acids were found. Based on the proton nmr studies, LMWCr appears to possess a tetranuclear assembly, although the presence of 2 dinuclear assemblies cannot be ruled out with certainty, perhaps reminiscent of the oxo-bridged, carboxylate-bridged dinuclear iron centers of non-heme iron proteins such as ribonucleotide reductase and the oxo-bridged, carboxylate-bridged tetramanganese assembly of photosystem II.

Chromium plays a crucial role in the activation of insulin receptor kinase activity by LMWCr. ApoLMWCr is inactive in the activation of the kinase activity in adipocytic membranes. However, titration of apoLMWCr with chromic ions results in the total restoration of the ability to activate kinase activity; approximately 4 chromic ions per oligopeptide were required for maximal activity, consistent with the presence of 4 chromium per molecule for isolated holoLMWCr. The reconstitution of the activation potential of LMWCr is specific to chromium. Of more than 3,000 citations about chromium published [26] since 1990, and for the general public, current data do not seem to warrant routine use of chromium supplements, as its risk–benefit ratio has not yet been adequately characterized.

References

1. Sawyer HJ (1994) Chromium and its compounds. In: Zenz C, Dickerson OB, Horvath EP (eds) Occupational medicine. Mo: Mosby Book Inc, St Louis
2. Snow ET (1994) Environ Health Perspect 102(3):41–44
3. Hathaway JA (1989) Sci Total Environ 86:169–179
4. Hendler SS (1990) The doctors' vitamin and mineral encyclopedia. NY simon and schuster 125–127
5. Cupo DY, Wetterhahn KE (1985) Cancer Res 45:1146–1151
6. Stearns DM, Wise JP, Patierno SR, Wetterhahn KE (1995) FASEB J 9:1643–1649
7. Sterns DM, Belbruno JJ, Wetterhahn KE (1995) FASEB J 9:1650–1657
8. Norseth T (1986) Br J Ind Med 43:649–651
9. Petrilli FL, DeFlora S (1987) Br J Ind Med 44:355
10. Levy LS, Martin PA, Venitt S (1987) Br J Ind Med 44:355–357
11. Mooradian A, Failla M, Hoogwerf B, Maryniuk M, Wylie-Rosett J (1994) Diabetes Care 17:464–479
12. Mertz W (1993) J Nutr 123:626–633
13. Anderson R, Polansky M, Bryden N, Canary J (1991) Am J Clin Nutr 54:909–916
14. Anderson RA, Cheng N, Bryden NA et al (1997) Diabetes 46:1786
15. Kapil V, Keogh J (1990) Chromium toxicity: case studies in environmental medicine. In: Rockville Md, (ed) Agency for toxic substances and disease registry, US Dept of Health and Human Services
16. Lim TH, Sargent T, Kusubov N (1983) Am J Physiol 244:R445–R454
17. Lee N, Reasner C (1994) Diabetes Care 17:1449–1452
18. American Diabetes Association Position statement (1996) Diabetes Care 19:S16–S19
19. Davis CM, Vincent JB (1997) Biochemistry 36:4382–4385
20. Davis CM, Sumrall KH, Vincent JB (1996) Biochemistry 35:12963–12969
21. Davis CM, Vincent JB (1997) J Biol Inorg Chem 2:675–679
22. Sumrall KH, Vincent JB (1997) Polyhedron 16:4171–4177
23. Davis CM, Vincent JB (1997) Arch Biochem Biophys 339:335–343
24. Davis CM, Royer AC, Vincent JB (1997) Inorg Chem 36:5316–5320
25. Yamamoto A, Wada O, Ono T (1984) J Inorg Biochem 22:91–102
26. Wada O, Wu GY, Yamamoto A, Manabe S, Ono T (1983) Environ Res 32:228–239